171 Topics in Current Chemistry

Electronic and Vibronic Spectra of Transition Metal Complexes I

Editor: H. Yersin

With contributions by
G. Blasse, A. Ceulemans, M. G. Colombo,
H. U. Güdel, A. Hauser, P. E. Hoggard,
C. Reber, H.-H. Schmidtke, D. Wexler,
J. I. Zink

With 76 Figures and 24 Tables

Springer-Verlag Berlin Heidelberg GmbH

This series presents critical reviews of the present position and future trends in modern chemical research. It is addressed to all research and industrial chemists who wish to keep abreast of advances in their subject.

As a rule, contributions are specially commissioned. The editors and publishers will, however, always be pleased to receive suggestions and supplementary information. Papers are accepted for "Topics in Current Chemistry" in English.

ISBN 978-3-662-14893-8 ISBN 978-3-540-48464-6 (eBook)
DOI 10.1007/978-3-540-48464-6

Library of Congress Catalog Card Number 74-644622

Originally published by Springer-Verlag Berlin Heidelberg New York in 1994
Softcover reprint of the hardcover 1st edition 1994

Typesetting: Macmillan India Ltd., Bangalore-25
Offsetprinting: Saladruck, Berlin; Bookbinding: Lüderitz & Bauer, Berlin
SPIN: 10128614 51/3020 - 5 4 3 2 1 0 - Printed on acid-free paper

Guest Editor

Prof. Dr. *Hartmut Yersin*
Institut für Physikalische und
Theoretische Chemie
Universität Regensburg
Universitätsstraße 31
93053 Regensburg, FRG

Editorial Board

Attention
all "Topics in Current Chemistry" readers:

A file with the complete volume indexes Vols.22 (1972) through 170 (1994) in delimited ASCII format is available for downloading at no charge from the Springer EARN mailbox. Delimited ASCII format can be imported into most databanks.

The file has been compressed using the popular shareware program "PKZIP" (Trademark of PKware Inc., PKZIP is available from most BBS and shareware distributors).

This file is distributed without any expressed or implied warranty.

To receive this file send an e-mail message to:
SVSERV@VAX.NTP. SPRINGER.DE
The message must be:"GET/CHEMISTRY/TCC_CONT.ZIP".

SVSERV is an automatic data distribution system. It responds to your message. The following commands are available:

HELP	returns a detailed instruction set for the use of SVSERV
DIR (name)	returns a list of files available in the directory "name",
INDEX (name)	same as "DIR",
CD <name>	changes to directory "name",
SEND <filename>	invokes a message with the file "filename",
GET <filename>	same as "SEND".

For more information send a message to:
INTERNET:STUMPE@SPINT. COMPUSERVE.COM

Preface

In recent years there has been a considerable amount of research on transition metal complexes due to the large number of potential or already realized technical applications such as solar energy conversion through photo-redox processes, optical information and storage systems, photolithographic processes, etc. Moreover, metal complexes are also of considerable importance in biology and medicine. Most of these applications are directly related to the electronic and vibronic properties of the ground and lowest excited states.

There is substantial practical and also scientific interest in developing a better understanding of the important electronic and vibronic structures of transition metal complexes than is presently available. Such knowledge is required for a more successful tailoring of complexes with specific, user-defined properties (e.g. extinction coefficients, emission wavelengths, lifetimes and quantum yields, spatial charge redistributions in excited states compared to ground states, photo-redox properties, etc.). For example, this appears feasible for transition metal complexes with organic ligands, for which the lowest excited electronic states can often be chemically tuned to possess mainly metal-centered (MC), metal-to-ligand-charge-transfer (MLCT), ligand-to-metal-charge-transfer (LMCT), ligand-to-ligand-charge-transfer (LL'CT), or ligand-centered (LC) character. Such controllable chemical variations can thereby lead to complexes with quite different properties.

Spectroscopic methods can yield the required understanding of the complexes. Especially optical spectroscopy provides very detailed information about electronic and vibronic structures, in particular, when highly resolved spectra are available. However, without the development of suitable models, which are usually based on perturbation theory, group theory, and recently also on ab-initio calculations, a thorough understanding of the complexes is very difficult to achieve. In this volume and in a subsequent one some leading researchers will show that such a detailed description of transition metal complexes can indeed be successfully achieved.

This volume provides a survey of modern developments for the description of electronic and vibronic structures of transition metal complexes. Since this research area is investigated by both chemists and physicists, who work theoretically as well as experimentally, it is clear that building a bridge between these groups is difficult. Nevertheless, it would be highly desirable to accomplish this aim at least partially. Thus, in a theoretical contribution A. Ceulemans discusses, in a shell-theoretical view, the still fascinating properties of the low-lying electronic states of Cr^{3+} complexes applying a second quantization treatment. P. Hoggard applies the angular overlap model to interpret spectra of Cr^{3+} complexes. H.-H. Schmidtke discusses the vibronic Herzberg-Teller and Franck-Condon coupling schemes and their importance to the vibrational satellite structures of different types of electronic transitions in various metal complexes. D. Wexler, J. I. Zink, and Ch. Reber examine effects of vibrational coordinate coupling on optical spectra. In a mainly experimentally based description G. Blasse presents a summary of vibronic properties found for ions doped into various solid matrices. And M. G. Colombo, A. Hauser and H. U. Güdel investigate, in a spectroscopically based contribution, the interplay of ^{3}LC and ^{3}MLCT states with respect to optical properties of Rh^{3+} and Ir^{3+} complexes. I would like to thank the authors for their efforts.

In this field, the late Professor Dr. Günter Gliemann was deeply involved. He became known to the scientific community through a large number of important publications and his and H. L. Schläfer's excellent textbook "Basic Principles of Ligand Field Theory", which has appeared in German, English, and even in Chinese. As my "Doktorvater" Günter Gliemann initiated my interest in transition metal complexes. Subsequently, we enjoyed very fruitful scientific cooperation, which led to numerous common publications, in particular, about chromium(III) complexes and tetracyanoplatinates(II). Professor Günter Gliemann died too early, at the end of 1990 at the age of 58. This volume is dedicated to him.

Regensburg, June 1994 Hartmut Yersin

Table of Contents

Vibrational Structure in the Luminescence Spectra of Ions in Solids

G. Blasse

Debye Research Institute, Utrecht University, P. O. Box 80.000, 3508 TA Utrecht, The Netherlands

Table of Contents

The luminescence spectra of metal ions in solids show often vibrational structure at low temperature. This structure yields direct information on the interaction between the metal ion and the lattice. This interaction is often strongly dependent on the nature of the surroundings of the emitting ion. This paper reviews this important aspect using illustrative examples out of the recent literature. The types of ions to be discussed are the following: ions with $d^{10}s^2$ configuration, rare earth ions and transition metal ions.

Topics in Current Chemistry, Vol. 171

1 Introduction

The optical spectra of ions in solids have been studied intensively, in absorption as well as in emission. A good and recent survey of the theory illustrated by many examples has been given by Henderson and Imbusch [1].

These spectra exist, characteristically, of bands which may be very broad or very narrow, and which may show vibrational structure. Examples of very broad bands without any structure at all, not even at very low temperatures, are found in the spectra of the F centre (an electron trapped at a halide vacancy in the alkali halides), and the tungstate (WO_4^{2-}) group in $CaWO_4$. The spectral width may approach a value of 1 eV, and the Stokes shift of the emission band may be 2 eV.

The intraconfigurational transitions of the rare earth ions ($4f^n$) are examples of ions which, even in solids, show sharp lines in their spectra. The width is of the order of wavenumbers and is at 4.2 K usually determined by inhomogeneous broadening. These lines are true zero-phonon lines. The vibronic transitions belonging to these lines are weak and often overlooked.

In between these two extremes there are ions which show spectra with bands, the width of which ranges between a few hundred and a few thousand wavenumbers. At low temperature these bands show often vibrational structure. These spectra are of a considerable interest, since they reveal in a direct way information on the interaction between the ion and the host lattice.

If this interaction is very weak, the zero-phonon line dominates in the spectrum (like in the rare earth ions). If the interaction is very strong, the spectra contain only broad bands from which not much information can be obtained. These situations are known as the weak- and strong-coupling case, respectively. Vibrational structure of any importance is usually only observed for the intermediate-coupling case. This is, for example, encountered for transition metal ions and uranate complexes.

It is not so simple that certain ions show no structure in their spectra, whereas other do. The nature of the host lattice plays also an important role. This is illustrated in an impressive way by the Bi^{3+} ($6s^2$) ion. Depending on the host lattice its spectra may show narrow bands with vibrational structure or very broad and structureless bands, and the Stokes shift of the emission may vary from 1000 to 20000 cm^{-1} [2].

By dealing with metal ions with different configurations this paper will illustrate the dependence of the spectral band shape on the nature of the metal ion. This, however, is a well-known and reasonably well understood phenomenon. By dealing with a given ion in different host lattices we will illustrate how the spectral band width depends on the nature of the host lattice. This dependence is of a larger complexity.

The structure of this review is as follows. In Chap. 2 we will survey shortly and nonmathematically the theories in use to explain spectral band width and structure. In subsequent chapters we will deal with a couple of different metal ions which have been studied intensively in several host lattices. These are the following: s^2 (Chap. 3), $4f^n$ (the rare earth ions, Chap. 4), and d^n (the transition metal ions, Chap. 5, including oxo-d^0-complexes and uranate complexes). The review closes with a short concluding paragraph.

2 Theoretical Models

The simplest model to describe the absorption and emission spectra of a centre in a solid is the single-configurational-coordinate model [1–4]. This is shown in Fig. 1. The ground state of the centre is given by parabola *a*, the excited state by parabola *b*. The energy *E* is plotted vertically, the single configurational coordinate *Q* horizontally. The ground state minimum is at Q_0^a, the excited state minimum at Q_0^b, so that the offset is $Q_0^b - Q_0^a$. The energy difference between the minima is E_0^b. The ground state vibrational wavefunctions are $\Psi(a, \mathrm{n})$, those of the excited state $\Psi(b, n')$.

Consider a vibronic transition between the states $\Psi(b, n')$ and $\Psi(a, n)$. The transition dipole moment for this transition is given by

$$\langle \Psi(a, n) | \mu | \Psi(b, n') \rangle \qquad (1)$$

where μ is the operator involved. The optical transition from *b* to *a* consists of a number of vibronic transitions, characterized by *n* and *n'*. At low temperatures $n' = 0$. The $n' = 0$ to $n = 0$ transition is called the zero-phonon transition.

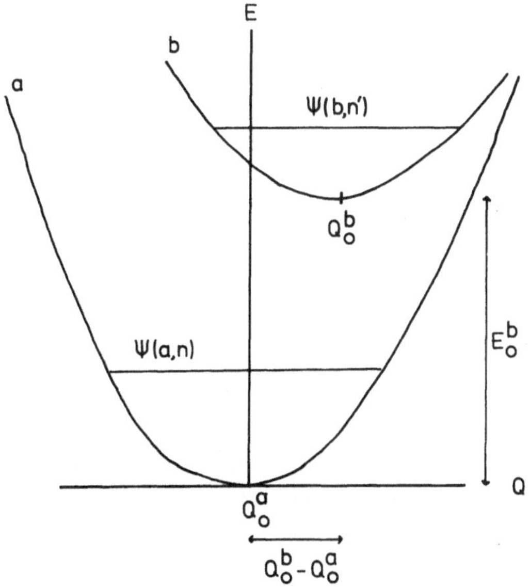

Fig. 1. The configurational coordinate diagram. The energy *E* is plotted versus a configurational coordinate Q. The offset between the parabolae is given by $Q_0^b - Q_0^a$. The ground state *a* contains vibrational levels with quantum number *n*, the excited state *b* with quantum number *n'*

At higher temperatures vibrational levels with $n' > 0$ are also occupied, so that the temperature dependence of the vibronic intensity in emission is given by $1 + \langle n' \rangle$, where $\langle n' \rangle$ is the average value of n' given by

$$\langle n' \rangle = (\exp(h\nu/kT) - 1)^{-1}. \tag{2}$$

The frequency ν relates to the vibrational mode involved.

The use of the Condon approximation simplifies Eq. (1) to

$$\langle \Psi(a)|\mu|\Psi(b)\rangle \langle \chi(n)|\chi(n')\rangle. \tag{3}$$

Here χ presents the vibrational wavefunctions.

Equation (3) consists of two matrix elements. The first one determines in how far the optical $b \rightarrow a$ transition is allowed or forbidden; the second one determines the shape of the $b \rightarrow a$ transition. For low temperatures (i.e. $n' = 0$) and equal force constants (i.e. equal vibrational frequencies in ground and excited states), the result for the square of the second matrix element is

$$\frac{e^{-S} \cdot S^n}{n!}, \tag{4}$$

where S is the Huang-Rhys or coupling parameter. For $S = 0$ the spectrum contains only the zero-phonon line (weak-coupling case). In the intermediate coupling case ($1 < S < 5$) the zero-phonon line is observable but followed by a series of vibronic lines with equal spacing $E = h\nu$. This is called a progression in the vibrational mode with frequency ν. The strongest line is one of the vibronic

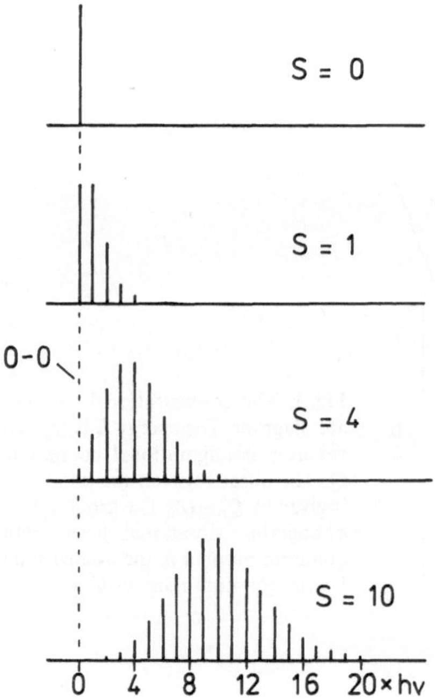

Fig. 2. Relative intensities of the electronic-vibrational lines for different values of the coupling parameter S. The envelope of the lines yields the bandshape

lines. At still larger values of S (strong-coupling case) the zero-phonon transition is not observable. These statements are illustrated in Fig. 2.

Since $\sum_n |\langle\chi(n)|\chi(0)\rangle|^2 = 1$, the intensity of the $b \rightarrow a$ transition does not depend on S; only its shape does. The intensity of the zero-phonon transition is a fraction e^{-s} of the total intensity.

The shift of the emission maximum relative to the absorption maximum, the so-called Stokes shift, is determined by the value of $Q_0^b - Q_0^a$ (see Fig. 1). For the equal force constant case this Stokes shift is equal to $2Sh\nu$ [2]. This indicates that the Stokes shift is small for the weak-coupling case and large for the strong-coupling case. It is also clear that the value of the Stokes shift, the shape of the optical bands involved, and the strength of the (electron-vibrational) coupling are related. For a more detailed account of these models the reader is referred to the literature mentioned above [1–4].

3 Luminescence of Ions with s^2 Configuration

3.1 General

The spectroscopy of ions with s^2 configuration in solids is variable and rich. Depending on the host lattice, and the charge and principal quantum number of the ion involved, the emission spectrum can consist of a broad structureless band or a narrow band with a considerable amount of vibrational structure [2]. The build-up of this structure depends on the principal quantum number. Here we will first review experimental data on $6s^2$ ions (like Pb^{2+} and Bi^{3+}) and on $4s^2$ and $5s^2$ ions (like Se^{4+} and Te^{4+}). After that we will compare these data and try to come to conclusions.

The energy level structure of an s^2 ion is simple [5, 6]. The s^2 ground state configuration yields one level, viz. $^1S_0$. The excited-state configuration, sp, yields the following levels in sequence of increasing energy: $^3P_0$, $^3P_1$, $^3P_2$ and $^1P_1$ (see

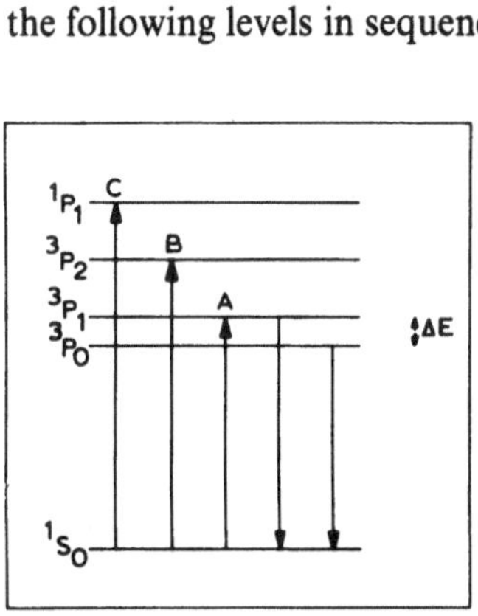

Fig. 3. Schematic representation of the energy-level scheme of a free s^2 ion

Fig. 3). Note that optical transitions between the two configurations are parity allowed as electric-dipole transitions. The spin-selection rule is relaxed by spin-orbit coupling, the more so the higher the principal quantum number is. Due to selection rules on ΔJ, the transitions $^1S_0$–$^3P_0$ and $^1S_0$–$^3P_2$ remain strongly forbidden. The emission is due to the $^3P_{0,1} \rightarrow {}^1S_0$ transition. Whether $^3P_0$ or $^3P_1$ is the initial level depends on their energy difference and the temperature.

3.2 Overview of Results

The Bi^{3+} ($6s^2$) ion has been investigated in many solids. The best model system is the elpasolite Cs_2NaYCl_6:Bi^{3+} [7] where the Bi^{3+} ion (on Y^{3+} sites) occupies an octahedral site with perfect cubic symmetry. Figure 4 presents the emission and excitation spectra of the luminescence of Cs_2NaYCl_6:Bi^{3+} at 5 K. These spectra relate to the $^1S_0$–$^3P_1$ transition. There is a clear progression in the breathing mode of the $BiCl_6^{3-}$ octahedron ν_1. This progression is built on the zero-phonon line and on the vibronic transitions due to coupling with a lattice mode and with the ν_2 and ν_5 of the $BiCl_6^{3-}$ octahedron [7]. See also Table 1. Figure 4 is, in a sense, a textbook example: it relates to an allowed transition, and the progression is in the breathing mode.

Such vibrational structure in the spectra of the Bi^{3+} ion has also been observed in a few other host lattices, viz. CaO and SrO (rock salt) [8, 9], $NaLnO_2$ (Ln=Sc, Y, Gd, Lu, ordered rock salt) [10], CaS [11], $YAl_3B_4O_{12}$ [12], $Ca_3(PO_4)_2$ [13], and $CaSO_4$ [14]. A structural requirement for the occurrence of this vibrational structure seems to be that the Bi^{3+} ion occupies a relatively small six-coordinated site, with $CaSO_4$ as an exception (eight co-ordination). Due to a low site symmetry and/or the simultaneous occurrence of

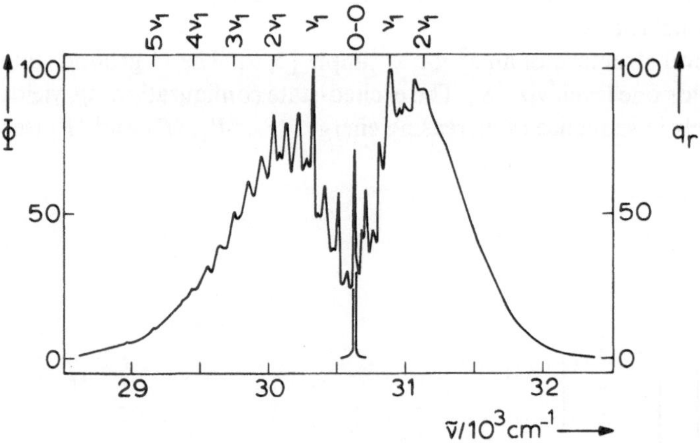

Fig. 4. The emission and excitation spectra of the luminescence of Cs_2NaYCl_6:Bi^{3+} at 5 K. The zero-phonon line (0–0) and the progression in the ν_1 mode are indicated at the top. See also Table 1. After A.C. van der Steen, thesis, Utrecht (1980)

the $^3P_0 \rightarrow {}^1S_0$ emission, the vibronic pattern is often more complicated than for Cs_2NaYCl_6:Bi^{3+}.

Usually, however, the Bi^{3+} emission spectrum does not show vibrational structure at all, not even at 4.2 K. An outstanding example is $Bi_4Ge_3O_{12}$ [15]. It shows a broad-band emission with an enormous Stokes shift (~2 eV). It is interesting to note that the Stokes shift of the Bi^{3+} emission varies from less than 1000 cm^{-1} (Cs_2NaYCl_6:Bi^{3+}) to 20000 cm^{-1} ($Bi_2Ge_3O_9$). This is illustrated in Table 2 [2]. A large Stokes shift implies a high value of the electron-lattice

Table 1. Vibrational structure in the spectra of Cs_2NaYCl_6:Bi^{3+} at 4.2 K (after Ref. [7]). All data in cm^{-1}. For 0–0 lines the actual spectral position is given, for vibronic lines the position relative to the corresponding 0–0 line

Emission $^3P_1 \rightarrow {}^1S_0$	$^3P_0 \rightarrow {}^1S_0$	Excitation $^1S_0 \rightarrow {}^3P_0$	Assignment*
30628	29304	30635	0–0
−46	−51	+47	ν_r
−112	−111	+78	ν_5
−144		+127	$\nu_5+\nu_r$
−200	−192	+185	ν_2
−258	−230		$\nu_2+\nu_r$
−294	−292	+259	ν_1
−343	−339	+309	$\nu_r+\nu_1$
−406	−402	+350	$\nu_5+\nu_1$
−444			$\nu_5+\nu_r+\nu_1$
−493	−476	+445	$\nu_2+\nu_1$
−540	−525		$\nu_2+\nu_r+\nu_1$
−593	−581	+500	$2\nu_1$
.	.		
.	.		
.	.		
.	.		

* 0–0: zero-phonon line; ν_r: lattice vibrations; $\nu_{1,2,5}$: octahedral $BiCl_6^{3-}$ vibrations.

Table 2. Some data on the Bi^{3+} luminescence in several host lattices (2)

Composition	Stokes shift of the emission (cm^{-1})	$\Delta E(^3P_1 - {}^3P_0)$ (cm^{-1})	Vibrational structure in the emission
Cs_2NaYCl_6:Bi	800	1150	yes
$ScBO_3$:Bi	1800	1000	yes
$YAl_3B_4O_{12}$:Bi	2700	1100	yes
CaO:Bi	2700	1200	yes
$LaBO_3$:Bi	8500	440	no
LaOCl:Bi	8500	540	no
La_2O_3:Bi	10800	370	no
$Bi_2Al_4O_9$	16000	25	no
$Bi_4Ge_3O_{12}$	17600	25	no
$LaPO_4$:Bi	19200	16	no
$Bi_2Ge_3O_9$	20000	16	no

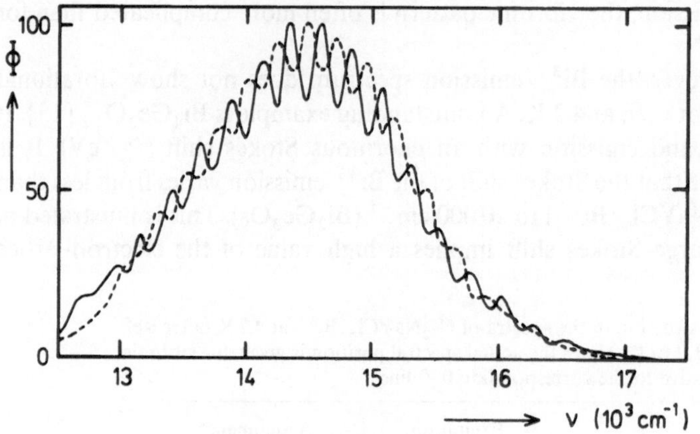

Fig. 5. The emission spectra of Cs_2ZrCl_6:Se^{4+} at 1.6 K (*dashed line*) and 4.2 K (*full line*). After H. Donker, thesis , Utrecht (1989). See also text

coupling parameter S and the absence of vibrational structure (see above). Also the parabolae offset $Q_0^b - Q_0^a$ will be large. It has been proposed that one reason for this is an off-centre position of Bi^{3+} in the ground state due to a pseudo Jahn-Teller effect, whereas it relaxes to the centre of the coordination polyhedron in the excited state. There are several experimental indications for such an off-centre position, obtained by X-ray diffraction and by EXAFS [2]. In solution similar observations have been made [16]. An accurate model of the situation is, however, lacking. It will be clear that for a Bi^{3+} ion in a small coordination polyhedron an off-centre position is not relevant due to space restrictions. This seems to explain the occurrence of vibrational structure reviewed above.

The Pb^{2+} ion is isoelectronic with the Bi^{3+} ion. The phenomena described above for the Bi^{3+} ion are also observed for the Pb^{2+} ion [2], but the occurrence of vibrational structure is even more restricted. It has been observed for CaO:Pb^{2+}, SrO:Pb^{2+} [9], and CaS:Pb^{2+} [17]. As far as we are aware no vibrational structure has ever been reported for the isoelectronic Tl^{+} ion, in spite of extensive studies on this ion [6, 9]. Therefore we conclude that the higher the charge of the $6s^2$ ion, the higher the probability to find vibrational structure in the emission spectrum.

The same seems to be true for the $4s^2$ and $5s^2$ ions, since vibrational structure has only been observed for Se^{4+} and Te^{4+}. These observations could only be made when these ions are in perfect-octahedral halide coordination. Examples are Cs_2MX_6 (M = Se, Te and X = Cl, Br) [18], A_2ZrCl_6:Te^{4+} (A = Cs, Rb) and A_2SnCl_6:Te^{4+} (A = Cs, Rb, K) [19], and A_2ZrCl_6:Se^{4+} (A = Cs, Rb) [20]. Observations of Te^{4+} luminescence in oxides are rare, and vibrational structure was absent (ZrP_2O_7:Te^{4+} [21] and tellurite anti-glasses [22]). Figure 5 presents, as an example of the vibrational patterns observed, the emission spectrum of Cs_2ZrCl_6:Se^{4+} at low temperature [20].

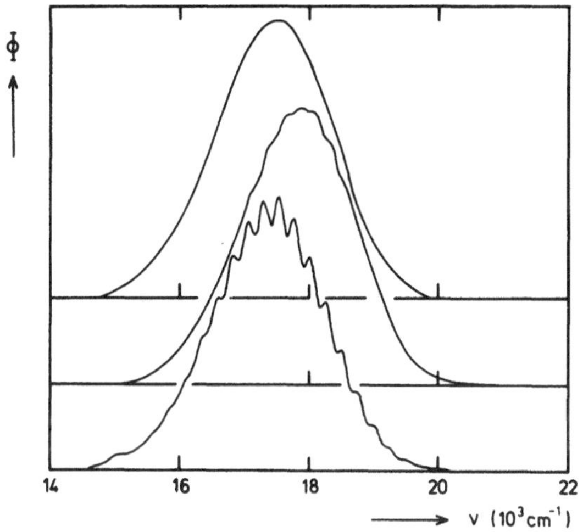

Fig. 6. The emission spectra of $A_2SnCl_6:Te^{4+}$ at 10 K. From top to bottom A = K, Rb, Cs. After H. Donker, thesis, Utrecht (1989)

This type of spectra consists of a long progression in the ν_2 vibrational mode of the SeX_6^{2-} or TeX_6^{2-} octahedron. The maximum of the emission band is situated at about the 10th member of the progression, i.e. $S \simeq 10$. Not only the mode on which the progression is based is different for $6s^2$ ions on one hand and $5s^2$ and $4s^2$ ions on the other (viz. ν_1 and ν_2, respectively), but also the value of S is much larger in the second case.

Actually the occurrence of vibrational structure for $5s^2$ and $4s^2$ ions depends critically on chemical composition. This is shown in Fig. 6 where the structure observed for $Cs_2SnCl_6:Te^{4+}$ is hardly visible in the emission spectrum of $Rb_2SnCl_6:Te^{4+}$, and has completely disappeared for $K_2SnCl_6:Te^{4+}$ [19].

3.3 Interpretation of Results

Now an approach will be made to explain the described results without going into too much detail. The latter has been done for specific cases, but here we aim at an understanding of the dependence of the emission on the natures of the s^2 ion and the host lattice in which it resides.

It is imperative to the discussion to realize that the excited ^3P state undergoes two types of interaction, viz. spin-orbit (SO) interaction and electron-lattice or Jahn-Teller (JT) interaction. The SO interaction will split the ^3P state into the $^3P_0$, $^3P_1$ and $^3P_2$ levels. The strength of this interaction increases with the nuclear charge, i.e. from $4s^2$ to $6s^2$ ions and from low to high ionic charge. The JT effect causes also a splitting of the ^3P state due to coupling of this state with vibrational modes. For octahedral coordination these are the e_g and t_{2g} (ν_2 and ν_5, respectively) modes of the MX_6 octahedron ($M = s^2$ ion, X = ligand). If the SO coupling is strong, the influence of the JT effect is expected to decrease [23].

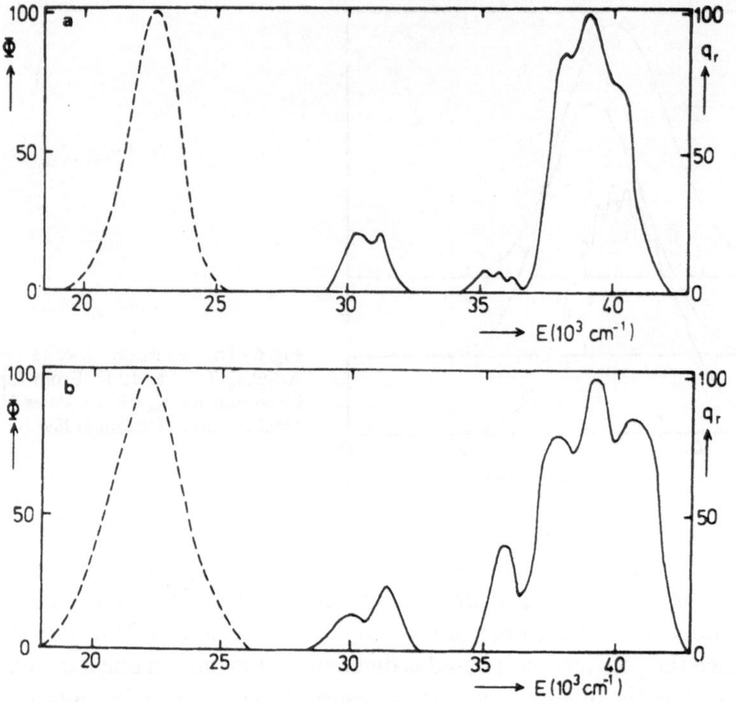

Fig. 7. Emission (*dashed lines*) and excitation (*full lines*) spectra of the luminescence of $Cs_2NaScCl_6$:Sb^{3+} at 4.2 K (**a**) and 300 K (**b**). After E.W.J.L. Oomen, thesis, Utrecht (1987)

The importance of the JT effect in the spectroscopy of s^2 ions can be easily observed from the splitting of the ${}^1S_0 \rightarrow {}^3P_1$, ${}^1P_1$ absorption (excitation) transitions [6]. Figure 7 gives as an example the spectra of Sb^{3+} ($5s^2$) in $Cs_2NaScCl_6$ at 4.2 and 300 K. The $SbCl_6^{3-}$ octahedron is cubic. The ${}^1S_0 - {}^3P_1$ transition at about 30000 cm^{-1} splits into two components, the ${}^1S_0 - {}^1P_1$ transition at about 40000 cm^{-1} into three components [24]. This, together with their temperature dependence, is proof of the dynamic JT effect in the ${}^3P$ state.

The consequences for the emission are even more drastic [6]. In some cases two emission bands are observed upon excitation in the ${}^1S_0 \rightarrow {}^3P_1$ transition. An example is given in Fig. 8, where the emission spectrum of YPO_4:Sb^{3+} is given as a function of temperature [25]. Clearly there are two emissions, one in the ultraviolet, the other in the blue spectral region. These two emission bands arise from different kinds of minima on the adiabatic potential energy surface (APES) of the ${}^3P$ relaxed excited state (RES). For cubic symmetry these minima are called the T (tetragonal symmetry) and X (trigonal symmetry) minima. In the case of octahedral coordination they are due to the interaction of the ${}^3P$ state with the e_g and t_{2g} vibrational modes, respectively [6, 26]. The relative positions of these two minima depend strongly on the ratio of the spin-orbit and the electron-phonon interactions, and on the site symmetry of the s^2 ion. Therefore

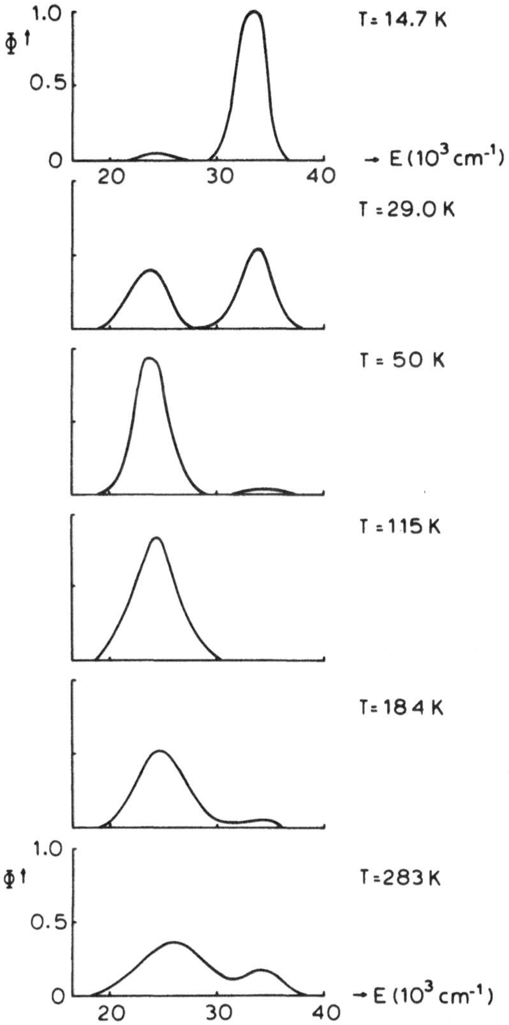

Fig. 8. Emission spectra of $YPO_4{:}Sb^{3+}$ as a function of temperature. After E.W.J.L. Oomen, thesis, Utrecht (1987)

some systems give only T emission, some others only X emission, and a few both types of emission [6].

The vibrational progression in the Se^{4+} and Te^{4+} emission bands indicates clearly the Jahn-Teller effect in the excited state: the coupling with the e_g vibrational mode points to a tetragonal distortion of the excited state, According to Schmidtke et al. [18], the magnitude of this tetragonal distortion is $\Delta z = -2\Delta x = -2\Delta y = 0.2$Å, a considerable distortion indeed.

Figure 6 shows that the occurrence of vibrational structure of this type is just on the boundary of what can be observed. Obviously it can be concluded, that no vibrational structure can be expected if not all factors which promote its occurrence are favourable. The difference between the spectra in Fig. 6 has been

Table 3. Stokes shift of the emission and spin-orbit coupling constant ζ of $5s^2$ ions in perovskite-like chlorides (after Ref. [19])

Composition	Stokes shift (cm^{-1})	ζ (cm^{-1})
$CsCaCl_3:Sn^{2+}$	15800	3000
$Cs_2NaYCl_6:Sb^{3+}$	8900	3800
$Cs_2ZrCl_6:Te^{4+}$	7400	4500

ascribed to slight deviations from cubic symmetry in the cases where the vibrational structure is obscured.

It is instructive to consider the Stokes shift of the emission. It depends on the coupling parameter S like the occurrence of vibrational structure does. Table 3 shows for three different $5s^2$ ions, Sn^{2+}, Sb^{3+} and Te^{4+}, in comparable host lattices the Stokes shift and the value of the spin-orbit coupling constant [19]. The luminescent ions have practically the same size as the host lattice ions involved. It is clear from the table that the Stokes shift increases drastically in the sequence of isoelectronic ions $Te^{4+} < Sb^{3+} < Sn^{2+}$.

Three contributions to the Stokes shift of an octahedrally coordinated s^2 ion have to be taken into account [19]:

1. a shift in the configurational coordinate diagram along the Jahn-Teller active coordinates e_g and t_{2g}. The vibrational progression observed in the Se^{4+} and Te^{4+} emissions shows that the shift along the e_g coordinate dominates.
2. a shift along the totally symmetric vibrational coordinate (a_{1g}).
3. a shift along vibrational coordinates of t_{1u} symmetry due to mixing of the electronic ground state with the excited states of T_{1u} symmetry (pseudo Jahn-Teller effect).

The increase of the Stokes shift in the sequence Te^{4+}, Sb^{3+}, Sn^{2+} is due to the first contribution. This is immediately clear by remembering that a strong spin-orbit coupling will quench the Jahn-Teller effect. Since Te^{4+} is the ion for which vibrational structure can just be observed, it is not surprising that the literature contains no examples of vibrational structure in the emission spectra of Sb^{3+} and Sn^{2+}.

In the tellurite antiglasses [22] the asymmetric coordination of Te^{4+} shows that the contribution c will also be important. In fact the Stokes shift is large (10000–17000 cm^{-1}) and vibrational structure is lacking.

In the $6s^2$ ions the contribution a will be weak due to the large spin-orbit coupling. If the contribution c is restricted due to a low coordination number and/or a small coordination polyhedron, only the contribution b remains. In fact, if the emission of $6s^2$ ions contains vibrational structure, it consists always of a progression in the a_{1g} vibrational mode. It is now also clear that this will be the easier observed if the charge is high, i.e. the spin-orbit coupling large, since this quenches the Jahn-Teller effect.

If $6s^2$ ions have asymmetric coordination, the contribution c seems to dominate. This follows from the enormous Stokes shift of the emission of

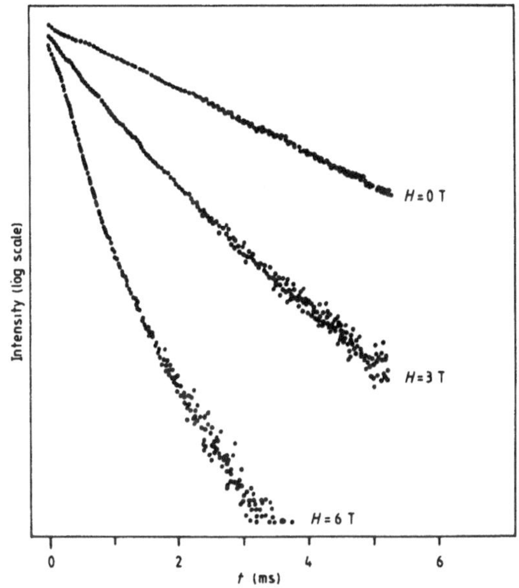

Fig. 9. Luminescence decay curves of Cs_2NaYCl_6:Sb^{3+} for different magnetic field strengths H. $T = 1.9$ K and H//<111>. Reproduced with permission from Ref. [29]

compositions like $Bi_4Ge_3O_{12}$ and $LaPO_4$:Bi^{3+} [2]. As we saw above for the tellurite antiglasses, this effect holds also for the other s^2 ions.

There exists also a relation between the Stokes shift and the energy difference ΔE between the $^3P_1$ and $^3P_0$ levels [27, 28]. These variables are inversely proportional if one compares a whole series of host lattices doped with s^2 ions. Large values of ΔE imply small Stokes shifts and should, therefore, be connected with the presence of vibrational structure. For Bi^{3+} this is convincingly the case (see Table 2).

We conclude that it is possible to account qualitatively for the presence or absence of vibrational structure in the emission of s^2 ions. In a volume dedicated to the memory of Prof. G. Gliemann this paragraph on s^2 ions should consider the contributions he made by the application of a magnetic field [29–31]. An applied magnetic field H shortens the luminescence decay. This effect depends on the relative orientation of the field and the crystal axes. Figure 9 shows as an example the decay of the Sb^{3+} emission of Cs_2NaYCl_6:Sb^{3+} at 1.9 K [29]. The shortening of the decay time by the magnetic field is due to magnetic-field-induced mixing of excited states; roughly speaking the $^3P_1$ state is admixed to the $^3P_0$ level. Further analysis reveals also the symmetry of the excited state. They are compatible with a tetragonal Jahn-Teller distortion owing to a strong coupling of the excited $^3P_1$ state and the e_g vibrational mode. In view of the vibrational progression in e_g, this is not unexpected. This has been shown for Cs_2NaMCl_6:Sb^{3+} (M = Sc, Y, La, Ref. [29] and $(NH_4)_2TeCl_6$ [30].

Gliemann et al. also studied the luminescence of systems with vibrational structure [31], viz. the Se^{4+} ion in Cs_2SeCl_6 and Rb_2SeCl_6. Figure 10 presents the emission spectrum of Cs_2SeCl_6 as a function of temperature and magnetic field. The low-temperature emission intensity increases with temperature as well

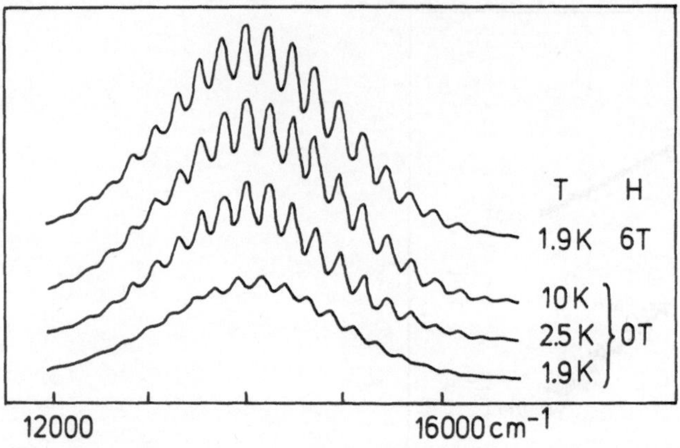

Fig. 10. Emission spectra of Cs_2SeCl_6 as a function of temperature and magnetic field strength. Reproduced with permission from Ref. [31]

as magnetic field (a factor of 3.5 at 1.9 K from $H=0$ to $H=6T$), whereas the decay time of the emission is shortened by a magnetic field (from 10 ms to 0.16 ms at 1.9 K from $H=0$ to $H=6T$). The interpretation is as follows:

The $SeCl_6^{2-}$ octahedron has perfect octahedral (O_h) symmetry in the ground state. The excited state is tetragonally distorted (D_{4h} symmetry) due to a static Jahn-Teller effect. As a consequence the $^3P_1$ state splits into $^3A_{2u}$ and ^3E_u. The spin degeneracy is removed by spin-orbit coupling: $^3A_{2u}$ yields a singlet A'_{1u} as the most stable RES and a doublet E'_u at higher energy.

An electric-dipole transition from E'_u to the ground state allowed; from A'_{1u}, however, it is forbidden. The latter transition can occur via vibronic coupling. At very low temperatures the emission is from A'_{1u}. With increasing temperature E'_u is occupied and the decay becomes faster, since $\Delta E(E_u'-A_{1u})$ is only 16 cm^{-1}. Actually the spectra in Fig. 10 for $T=1.9$ K and 2.5 K are different (compare also Fig. 5), since they belong to the A'_{1u} and E_u emission, respectively. At 2.2 K all structure has disappeared due to spectral overlap of the two emission bands.

The magnetic field H mixes the A'_{1u} and E'_u states. Therefore the decay of the A'_{1u} emission shortens and its intensity increases upon application of the field. The A'_{1u} emission is more allowed (i.e. it does not need vibronic assistance) and the emission pattern has changed (Fig. 10). From the orientation of the crystal in the magnetic field it can again be deduced that the emitting RES has tetragonal symmetry. Nevertheless, also this system leaves still problems for further study [20, 31].

4 Luminescence of Rare Earth Ions

The vibronic transitions in the spectra of the rare earth ions have recently been reviewed [32]. See also the chapter by Prof. C. D. Flint in this book. Here

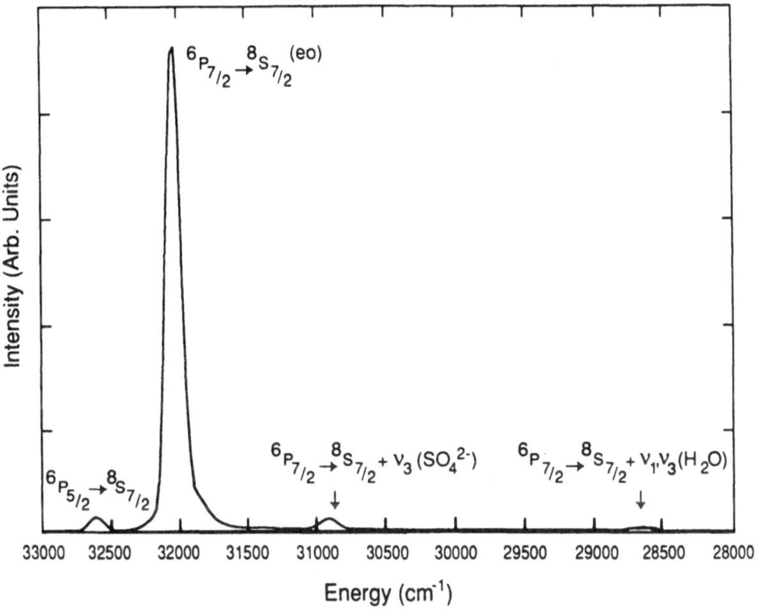

Fig. 11. X-ray excited emission spectrum of $Gd_2(SO_4)_3 \cdot 8H_2O$ at 300 K. See also Table 4. Reproduced with permission from Ref. [33]

we will mention only some of the main topics. Vibronic transitions have been observed for two completely different types of transitions, viz. the intraconfigurational $4f^n$ transitions and the interconfigurational $4f^n$–$4f^{n-1}5d$ transitions.

The intraconfigurational $4f^n$ transitions are strongly parity-forbidden and have $S \sim 0$. In good approximation the corresponding spectra show only the zero-phonon transitions. Vibronic transitions have integrated intensities which are a few percent of the corresponding zero-phonon transitions. In spite of their weakness these vibronic transitions yield useful information on the coupling strength between the $4f$ electrons and the surroundings of the rare-earth ion.

The Gd^{3+} ion ($4f^7$) is extremely suitable for vibronic studies since the electronic ground level, ^8S, is orbitally nondegenerate. As an example we take $Gd_2(SO_4)_3 \cdot 8H_2O$ [33]. The Gd^{3+} ion in this compound is coordinated by sulfate ions as well as by water molecules. Figure 11 gives the emission spectrum. The vibronic transitions due to coupling with the stretching modes of the SO_4^{2-} and the H_2O species are clearly observed. Table 4 gives the complete spectrum with an assignment.

Upon dehydrating this octahydrate, the vibronic transitions in which water vibrations are involved disappear. By hydrating $Gd_2(SO_4)_3$ with D_2O they reappear, but at the expected lower frequency (2440 cm^{-1} instead of 3330 cm^{-1}). This proves directly the role of the water molecule in this transition.

In some cases vibronic transitions due to coupling with vibrations in the second coordination sphere have been observed. An example is $(NH_4)_2GdCl_5$.

Table 4. The $^6P_{7/2} \rightarrow ^8S$ emission of Gd^{3+} in $Gd_2(SO_4)_3 \cdot 8H_2O$ at room temperature [33]

Position of emission line (cm^{-1})	Intensity[a]	Assignment[b]
32025	vs	0–0
~31850	w	ν_r
31390	w	ν_4 (SO_4^{2-})
30910	w	ν_3 (SO_4^{2-})
30420	vvw	ν_2 (H_2O)
28696	vw	ν_1, ν_3 (H_2O)

[a] v: very; w: weak; s: strong
[b] 0–0: zero-phonon transition; ν_r: lattice mode

In this compound there are $GdCl_6^{3-}$ octahedra in which the Gd^{3+} ion is shielded from the further surroundings by the large Cl^- ions. Nevertheless the photo-excited emission spectrum shows not only vibronic transitions due to coupling with the Gd–Cl vibrations, but also due to NH_4^+ bending and stretching vibrations [34].

Peculiarly enough, the rare-earth vibronic intensities depend on the concentration of the rare-earth ion [32]. From this effect an interaction range between the rare-earth ions of some 10 Å can be estimated. The nature of this interaction is not yet clear, but covalency seems to play a role.

The theory of these transitions is only in a developing state. What is known and in how far it can be applied to interpret experimental spectra has been reviewed in Ref. [32]. It appears that the vibronic intensities are for the greater part determined by covalency.

The interconfigurational $4f^n$–$4f^{n-1}5d$ transitions are of a different nature. They are parity allowed and can be described with the intermediate-coupling model, i.e. $1 < S < 5$. In some cases a beautiful vibrational structure with a short progression has been observed.

Figure 12 gives as an example the emission spectrum of $SrB_4O_7:Yb^{2+}(4f^{14})$. There is a clear progression in a frequency of 75 cm^{-1}, yielding $S = 6$ [35]. A rich vibrational structure corresponding to the Raman spectrum has been observed for $CaSO_4:Eu^{2+}$ ($4f^7$) at liquid helium temperatures [36, 37]. There is not much doubt that a stiff surroundings of the rare-earth ion in the host lattice promotes the occurrence of vibrational structure, and decreases the Stokes shift and the coupling parameter S. This is directly related to the fact that in this case there will be a relatively small offset of the parabolae ($Q_0^b - Q_0^a$) in the configurational coordinate model (Fig. 1) [2].

Edelstein et al. [38, 39] have studied $5f \rightarrow 6d$ transitions of some actinide ions ($5f^n$). The $5f \rightarrow 6d$ transition of Pa^{4+} ($5f^1$) in Cs_2ZrCl_6, for example, shows a nice vibrational structure, in absorption as well as in emission. The value of S amounts to about 2. The progression is in the symmetrical stretching (ν_1) mode of the $PaCl_6^{2-}$ octahedron. Interestingly enough, some weak features due to

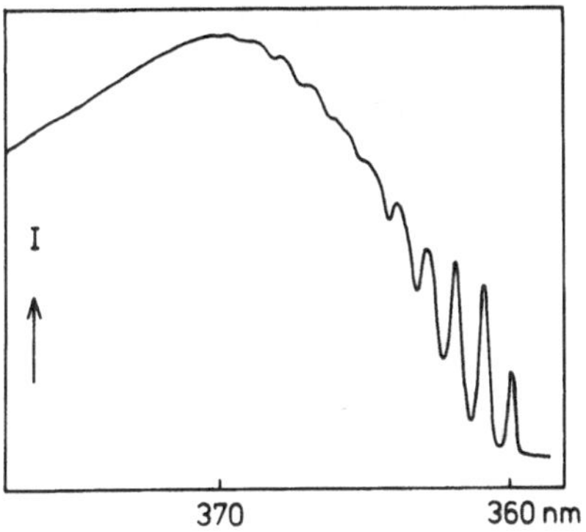

Fig. 12. Emission spectrum of $SrB_4O_7{:}Yb^{2+}$ at 4.2 K. Reproduced with permission from Ref. [35]

coupling with the Jahn-Teller-active modes of this octahedron are present in the emission spectrum, but not in the absorption spectrum. This is due to the fact that the potential energy surfaces for the Jahn-Teller-active modes are different in the $5f$ and the $6d$ states [38].

5 Luminescence of Transition Metal Ions

In this paragraph we will discuss two types of optical transitions in the transition metal ions: first the intraconfigurational d^n transitions, second the interconfigurational charge-transfer transitions in complexes with a "d^0" central metal ion.

The vibronic transitions in the intraconfigurational d^n (crystal-field) transitions have been discussed at length elsewhere (see e.g. Ref. [1]). We concentrate here on new emissions situated in the infrared spectral region and on information from vibronic spectra on distortions in the excited state.

Güdel and coworkers have reported during recent years many cases of (near) infrared emission from several transition-metal ions. This was only possible by the use of suitable detectors of radiation (e.g. a cooled germanium photodetector) and careful crystal synthesis. Here we mention some examples.

Compositions $CsMg_{1-x}Ni_xCl_3$ show Ni^{2+} emission at about 5000 cm^{-1} [40]. The emission band shows vibrational structure yielding an S value of about 2.5. From this value $Q_0^b - Q_0^a$ is found to be 0.7 Å, which gives $\Delta r = 0.24$ Å for the change in the Ni–Cl distance. This emission is due to a transition from one of the crystal-field components of the first excited state $^3T_{2g}$ ground state ($3d^8$, O_h notation). The lifetime of the excited state is 5.2 ms. The luminescence is quenched above 200 K.

In the analogous bromide system the values of S, $Q_0^b - Q_0^a$, and Δr are larger than in the chloride system; the Stokes shift is also larger, and the quenching temperature of the luminescence is lower, in agreement with the arguments given above.

The properties of V^{2+} ($3d^3$) were investigated among others in $MgCl_2$ [41]. The emission is due to the $^4T_2 \rightarrow {}^4A_2$ transition and is situated at about 7000 cm^1. The vibrational structure yields S values of about 5. The excited state appears to be distorted due to the Jahn-Teller effect. Above 250 K the emission is quenched.

Impressive is a study of the Ti^{2+} ion in $MgCl_2$ [42, 43]. The infrared emission is due to the $^1T_{2g} \rightarrow {}^3T_{1g}$ transition (O_h notation) around 7000 cm^{-1}, but there is also visible emission from the $^3T_{1g}$ level. The spin-forbidden transition shows an analogy with the ruby R lines. The lifetime of the emitting state is the longest-lived d-d luminescence (viz. 109 ms!). This is due to the weak spin-orbit coupling. The orbital degeneracy of the ground state leads to a great deal of structure in the emission spectrum.

By codoping $MgCl_2$ with Ti^{2+} and Mn^{2+} the authors were also able to study clusters like $Ti^{2+}Mn^{2+}$ and $Mn^{2+}Ti^{2+}Mn^{2+}$. This is possible by applying site-selective dye-laser spectroscopy. The exchange interactions in these clusters are considerable. The long lifetime mentioned above is reduced by two orders of magnitude (exchange induced intensity in the singlet-triplet transition).

In connection with transition-metal infrared emission, a recent report of Cu^{2+} luminescence must be mentioned. Dubicki et al. [44] have investigated crystals with composition $KCu_{0.01}Zn_{0.99}F_3$ and $K_2Cu_{0.01}Zn_{0.99}F_4$ and observed $^2T_{2g} \rightarrow {}^2E_g$ Cu^{2+} emission with magnetic-dipole zero-phonon lines at 6830 and 7498 cm^{-1}, respectively. The low-temperature lifetimes are of the order of 1 μs. It turns out that the spin-orbit coupling quenches the Jahn-Teller coupling in the excited state, so that the emitting $^2T_{2g}$ state has octahedral geometry. The final state is the lower level of the Jahn-Teller split ^2E_g state.

A much higher charged ion like Mn^{5+} also shows emission [45]. Due to the high crystal field, this emission is due to the $^1E \rightarrow {}^3A_2$ transition. Figure 13 gives spectra for Mn^{5+} in $Ba_5(PO_4)_3Cl$ and Ca_2VO_4Cl. We are clearly dealing with the weak-coupling case. In addition to the prominent zero-phonon transition there are a large number of vibronic lines. The assignment in terms of Mn–O bending (δ) and stretching (ν) vibrations is indicated in the figure. The progression is in the bending vibration, indicating a distortion in the excited state. In view of the intensity distribution S is small, and the distortion also.

Extremely interesting results were obtained for Mn^{4+} in Cs_2GeF_6:Mn^{4+} by McClure et al. by using two-photon spectroscopy [46]. Although this involves only absorption spectroscopy, the results are mentioned here because they are of interest for the understanding of ion-lattice coupling. The MnF_6^{2-} octahedron in Cs_2GeF_6 has perfect octahedral symmetry. The transitions within the $3d^3$ shell are parity forbidden for one-photon spectroscopy, resulting in very weak zero-phonon lines and strong vibronics due to coupling with ungerade vibrations (false origins). In two-photon spectroscopy the zero-phonon lines are allowed and are expected to dominate. This makes a complete analysis of the vibrational structure possible without complications. Here we mention only a few results.

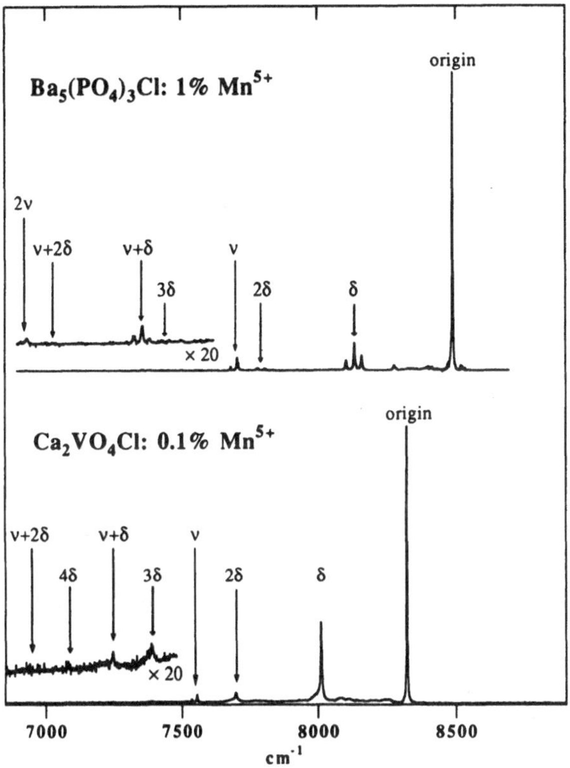

Fig. 13. Emission spectra of $Ba_5(PO_4)_3Cl{:}Mn^{5+}$ and $Ca_2VO_4Cl{:}Mn^{5+}$ at 8 K. The zero-phonon transition is the dominating line on the right-hand side. The vibronic lines are assigned in terms of the Mn–O bending (δ) and stretching (ν) vibrations. Reproduced with permission from Ref. [45]

The ${}^4A_2 \rightarrow {}^2E$ transition (the reverse of the emission transition) shows that the expansion in the ${}^2E$ level is only 0.003 Å (weak-coupling scheme). The ${}^4A_2 \rightarrow {}^4T_2$ transition shows that for the ${}^4T_2$ level this expansion is much larger (viz. 0.053 Å), and $S=3$ (intermediate coupling scheme). However, there is not only coupling with the $\nu_1(a_{1g})$ mode, but also with e_g and t_{2g} (Jahn-Teller active modes). This is even more pronounced in the ${}^4A_2 \rightarrow a^4T_{1g}$ transition, which shows a clear progression in e_g with a maximum intensity at the fifth member of the progression. The corresponding value of S for this mode is 5. This shows that in the excited a^4T_{1g} state there occurs a larger Jahn-Teller distortion (see below).

During the last decade it has been observed frequently from emission spectra that the excited state is distorted by the Jahn-Teller effect. A detailed study is that by Güdel et al. [47] on Cr^{3+} in elpasolites ($Cs_2NaInCl_6$, Cs_2NaYCl_6 and Cs_2NaYBr_6). The former two show broadband emission with vibrational structure. There are progressions in $\nu_1(a_{1g})$ and in $\nu_2(e_g)$. This enables us to obtain an accurate picture of the relaxed excited ${}^4T_{2g}$ level. Table 5 gives S values derived from the progressions in a_{1g} and e_g and from the Stokes shift. We are clearly dealing with an intermediate coupling case. Table 5 also gives the

Table 5. Parameters characterizing the $^4T_2 \rightarrow ^4A_2$ emission of Cr^{3+} in some cubic elpasolites[a]

Host lattice	$Cs_2NaInCl_6$	Cs_2NaYCl_6	Cs_2NaYBr_6
$S\ (a_{1g})$[b]	1.6	1.6	[c]
$S\ (e_g)$[b]	1.1	1.7	[c]
S_{total}[d]	3.7	3.6	6.7
Stokes shift (cm^{-1})	1780	1740	2200
$\Delta Q(a_{1g})$ (Å)	0.13	0.13	0.22[e]
$\Delta Q(e_g)$ (Å)	−0.13	−0.15	−0.22[e]
$\Delta x, y$ (Å)	0.09	0.10	0.15
Δ (Å)	−0.02	−0.03	−0.05

[a] Data from Ref. [47]
[b] From intensity ratios in vibrational progression
[c] No vibrational structure observed
[d] From the value of the Stokes shift
[e] Assuming $\Delta Q(a_{1g}) = \Delta Q(e_g)$, where $\Delta Q = Q_0^b - Q_0^a$ (see Fig. 1).

distortions. We observe again that the distortions are larger if the host-lattice ion involved is larger ($r_{Y^{3+}} > r_{In^{3+}}$) or the anion is larger ($r_{Br^-} > r_{Cl^-}$). The e_g progression is a clear indication of a Jahn-Teller effect in the $^4T_{2g}$ level.

In connection with our treatment of the configurational coordinate diagram (see above) it should be noticed that the occurrence of two progressions for Cr^{3+} in the elpasolites indicates that a multi-configurational coordinate diagram must be used.

Let us now turn to the charge-transfer transitions. These are well known from the efficient luminescence of complexes like VO_4^{3-} (in YPO_4), NbO_4^{3-} (in $YNbO_4$), WO_4^{2-} (in $CaWO_4$), and WO_6^{6-} (in Ba_2CaTeO_6). A review of this type of luminescence has been given elsewhere [48]. However, all these emissions are of a broad-band type (half width $\geq 5000\ cm^{-1}$) with large Stokes shifts, so that they belong to the strong-coupling case. Vibrational structure is not observed, not even at the lowest temperatures.

It has long been realized that the transitions involved are of the charge-transfer type, and that of emitting state is a triplet state [48]. This follows from the long decay times observed. Further characterization is hardly possible, due to the absence of any structure in the spectra. By performing EPR measurements in the excited state Van der Waals et al. [49–53] were able to demonstrate directly the triplet character of this excited state.

Let us consider $CaMoO_4$ [51]. In the ground state there is one MoO_4^{2-} group with S_4 site symmetry in the unit cell. However, in the excited state there are four magnetically equivalent MoO_4^{2-} groups in the unit cell. These show a strong trigonal distortion due to a static Jahn-Teller effect. This distortion can occur along any of the four Mo–O bonds. This means that the configurational coordinate diagram should take a Jahn-Teller active mode into account (probably the ν_2 mode). It is interesting to note that the interactions entering the

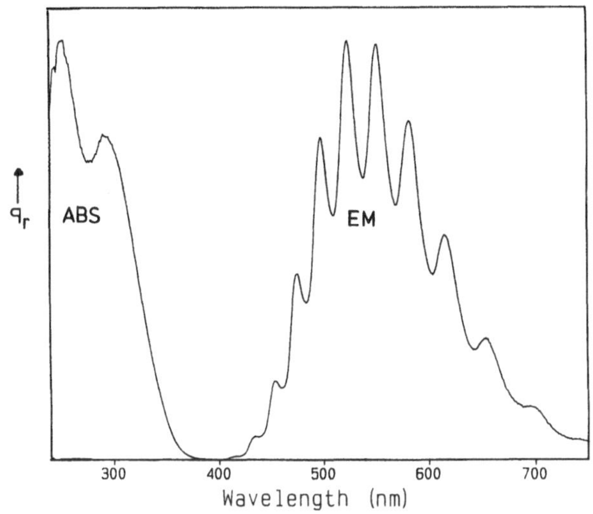

Fig. 14. The emission spectrum of the silica-supported vanadate group at 4.2 K. After M.F. Hazenkamp, thesis, Utrecht (1992)

problem decreases in the order: Jahn-Teller coupling > crystal field > spin-orbit coupling.

Very similar results have been observed for YVO_4 (with a VO_4^{3-} group). In $Ba_3(VO_4)_2$ the site symmetry is C_{3v}. Therefore only one or three inequivalent excited species are expected. The latter possibility turned out to be the case. This confirms the model sketched [52].

In recent years, however, it has become clear that some of these complexes show vibrational structure. The very first report was the low-temperature emission spectrum of $K_2Cr_2O_7$ [54]. This is one of the rare chromates which show luminescence. The emission spectrum shows a progression in 361 cm^{-1}, which is clearly a bending vibration of the chromate complex.

Less well-defined is the vanadate complex on a silica surface [55–58]. This complex shows an emission band with a clear progression in a vibrational mode with frequency of 1050 cm^{-1} (see Fig. 14). This vibration has to be assigned to a surface vanadyl group of a tetrahedral vanadate group, i.e. three V–O bonds are linked to the silica and one points out of the surface. The most intense member of the progression is the fifth. Analysis of the intensity distribution in the progression yields that the excited state has expanded 0.12 Å relative to the ground state. This plays an important role in photoreactions using SiO_2–V as a catalyst [57].

The decay time of this emission is very long, viz. some 5 ms [57, 58]. There are two reasons for this. First the transition involved is spin forbidden [48, 51]; secondly, the spin-allowed transition from which the spin-forbidden transition steals its intensity is unusually weak [58, 59].

For $KVOF_4$ very similar results have been obtained [59]. In the crystal structure of this solid there are isolated VOF_4 groups which contain a vanadyl group. Ab-initio calculations on complexes with a short V–O bond can account for the low oscillator strength of the lowest absorption transition, and therefore also of the emission transition [60, 61].

For chromate on silica such a vibrational structure in a high-frequency stretching mode has been observed too [62]. In solid $K_2NbOF_5 \cdot H_2O$, however, the $NbOF_5$ complex with a short Nb–O bond shows an emission spectrum with a progression in a Nb–F deformation mode [59]. A very similar emission spectrum was reported for $K_2NaTiOF_5$ which contains $TiOF_5$ octahedra with a titanyl group which is responsible for the optical transitions. Due to the high symmetry in this lattice a precise assignment was possible. There is a long progression in a frequency of 290 cm^{-1}, the vibronic line with the highest intensity being the 15th member of the progression. In this vibration the central titanium ion moves relative to the horizontal plane of fluoride ions (Fig. 15). Excitation of the Ti–O group will result in an increase of the Ti–O distance, so that the Ti^{4+} ion moves relative to the plane of the four F^- ions. In this way the charge-transfer state may be coupled with the TiF_4 vibration. There was no vibronic line observed due to coupling with the Ti–O stretching vibration, but it could easily go unnoticed under the progression in the bending vibration.

From these observations it can be concluded that the excited state of the oxo-d^0-complexes is generally distorted. Only the linear complexes show in some cases a progression in the stretching vibration which is obvious. It should be noted that in practically all cases the progressions in this type of emission spectra are on the boundary of what can be observed. Results are usually inferior to the observations depicted in Fig. 14.

In closing this paragraph we mention the emission spectroscopy of the U^{6+} ion. This ion shows emission spectra with clear vibrational structure. It is mentioned here because the optical transition is of the charge-transfer type, and the U^{6+} ion may be described in the present context as a $5f^0$ ion. The vibrational structure in the emission spectra depends strongly on the ligands of

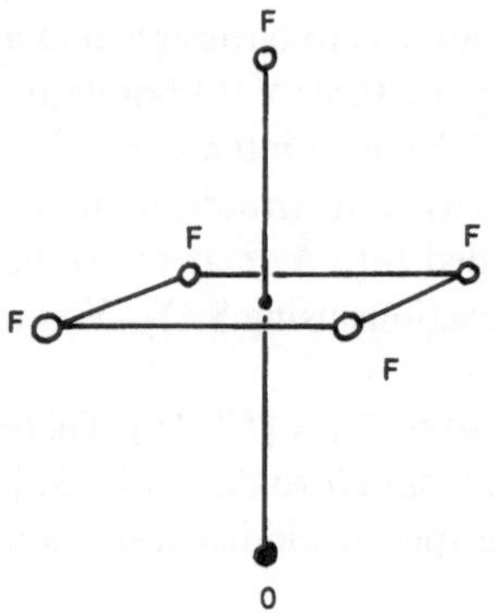

Fig. 15. The $TiOF_5$ octahedron in $K_2NaTiOF_5$. Reproduced with permission from Ref. [63]

the U^{6+} ion. In uranyl (UO_2^{2+}), for example, the progression, which is in the symmetrical stretching vibration, is longer than in complexes like UO_6^6, or UO_4^2 [64].

A special case in the uranyl spectroscopy is formed by Denning's extensive work on the UO_2^{2+} group in $Cs_2UO_2Cl_4C$ ([65], and papers cited therein). Later this work was extended to two-photon spectroscopy, a powerful method for a complex with inversion symmetry [66]. Excited state spectroscopy followed soon after [67]. The latter experiments show a progression in the excited-state absorption spectrum of the UO_2^{2+} complex in a frequency of 585 cm^{-1}. This is the U–O stretching frequency in the second excited state. In the first excited state it is at 715 cm^{-1}. In this way it is possible to determine that the U–O distance expands from 1.77 Å in the ground state to 1.84 Å in the first and 1.95 Å in the second excited state. This work has been reviewed recently [68].

6 Concluding Remarks

The presence of vibronic structure in the emission spectra of ions in solids does not depend on the nature of the particular ion alone, but also on the nature of its surroundings. In the case of ns^2 ions the strength of the spin-orbit interaction relative to the Jahn-Teller effect determines whether the progression will be in ν_1 or ν_2. However, it is not usually possible to observe any vibronic structure at all due to deviations from high symmetry (pseudo Jahn-Teller effect in the ground state). Our present understanding is of a qualitative nature. Further progress is hampered by the fact that the presence of vibronic structure in the spectra is for these ions more exception that rule.

The problems in case of the rare earth ions, a weak-coupling case, are completely different. Progress has only started a couple of years ago and will certainly accelerate in the coming years.

Crystal-field spectroscopy of d^n ions is being extended nowadays to the near-infrared region with interesting results. Vibronic structure is nowadays used to obtain information on the deformation of the excited state. Also in case of the closed-shell d^0 complexes the excited state appears to be strongly distorted. Among the latter class especially the linear species show efficient luminescence. Molecular-orbital calculations are in progress and will probably yield interesting results.

Vibronic emission spectroscopy of ions in solids is, therefore, of a very broad nature and is a field which is still expanding. Although the overall pattern in the interpretation of the results is nowadays clear, much work has still to be done in order to obtain a complete understanding of the vibronic structure in the emission spectra of the different types of ions in solids.

7 References

1. Henderson B, Imbusch GF (1989) Optical spectroscopy of inorganic solids, Clarendon, Oxford
2. Blasse G (1988) Progress Solid State Chem 18: 79
3. Struck CW, Fonger WH (1991) Understanding luminescence spectra and efficiency using W_p and related functions, Springer, Berlin Heidelberg New York
4. Stoneham AM (1985) Theory of defects in solids, Clarendon, Oxford
5. Seitz F (1938) J Chem Phys 6: 150
6. Ranfagni A, Mugni D, Bacci M, Viliani G, Fontana MP (1983) Adv Physics 32: 823
7. van der Steen AC (1980) Phys Stat Sol (b) 100: 603
8. Hughes AE, Pells GP (1975) Phys Stat Sol (b) 71: 707
9. van der Steen AC, Dijcks LTF (1991) Phys Stat Sol (b) 104: 283
10. van der Steen AC, van Hesteren JJA, Slok AP (1981) J Electrochem Soc 128: 1327
11. Yamashita N, Asano S (1976) J Phys Soc Japan 40: 144
12. Kellendonk F, van Os MA, Blasse G (1979) Chem Phys Letters 61: 239
13. van der Steen AC, Aalders Th JA (1981) Phys Stat Sol (b) 103: 803
14. van der Voort, Blasse G (1992) J Solid State Chem 99: 404
15. Rogemond F, Pédrini C, Moine B, Boulon G (1985) J Luminescence 33: 455
16. Nikol H, Vogler A (1991) J Am Chem Soc 113: 8988.
17. Asano S, Yamashita N (!978) Phys Stat Sol (b) 89: 663
18. Wernicke R, Kupta H, Ensslin W, Schmidtke HH (1980) Chem Phys 47: 235
19. Donker H, Smit WMA, Blasse G (1989) J Phys Chem Solids 50: 603
20. Donker H, van Schaik W, Smit WMA, Blasse G (1989) Chem Phys Letters 158: 509
21. Donker H, den Exter MJ, Smit WMA, Blasse G (1989) J Solid State Chem 83: 361
22. Blasse G, Dirksen GJ, Oomen EWJL, Trömel M (1986) J Solid State Chem 63: 148
23. Ham FS (1965) Phys Rev 132: 1727
24. Oomen EWJL, Smit WMA, Blasse G (1986) J Phys C: Solid State Phys 19: 3263
25. Oomen EWJL, Smit WMA, Blasse G (1988) Phys Rev B 37: 18
26. Fukuda (1970) Phys Rev B1: 4161
27. Blasse G, van der Steen AC (1979) Solid State Comm 31: 993
28. Mugnai D, Ranfagni A, Pilla O, Vilini G, Montagna M (1980) Solid State Comm. 35: 975
29. Meidenbauer K, Gliemann G, Oomen EWJL, Blasse G (1988) J Phys C: Solid State Phys 21: 4703
30. Meidenbauer K, Gliemann G (1988) Z Naturforschung 43a: 555.
31. Hesse K, Gliemann G (1991) J Phys Chem 95: 95
32. Blasse G (1992) Int Revs Phys Chem 11: 71
33. Brixner LH, Crawford MK, Blasse G (1990) J Solid State Chem 85: 1
34. Blasse G, Dirksen GJ (1992) J Solid State Chem 96: 258
35. Blasse G, Dirksen GJ, Meijerink A (1990) Chem Phys Letters 167: 41
36. Ryan FM, Lehmann W, Feldmann DW, Murphy J (1974) J Electrochem Soc 121: 1475
37. Yamashita N, Yamamoto I, Ninagawa K, Wada T, Yamashita Y, Nakov Y (1985) Jap J Appl Phys 24: 1174
38. Edelstein N, Kot WK, Krupa JC (1992) J Chem Phys 96: 1
39. Piekler D, Kot WK, Edelstein N (1991) J Chem Phys 94: 942
40. Reber C, Güdel HU (1986) Inorg Chem 25: 1196
41. Gälli B, Hauser A„ Güdel HU (1985) Inorg Chem 24: 2271
42. Jacobsen SM, Smith WE, Reber C, Güdel HU (1986) J Chem Phys 84: 5205; Jacobsen SM, Güdel HU (1989) J Luminescence 43: 125
43. Herren M, Jacobsen SM, Güdel HU, Briat B (1989) J Chem Phys 90: 663
44. Dubicki L, Kramer E, Riley M, Yamada I (1989) Chem Phys Letters 157: 315
45. Herren M, Güdel HU, Albrecht C, Reinen D (1991) Chem Phys Letters 183: 98
46. Chien RL, Berg JM, McClure DS, Rabinowitz P, Perry BN (1986) J Chem Phys 84: 4168
47. Knochenmuss R, Reber C, Rajasekharan MV, Güdel HU (1986) J Chem Phys 85: 4280
48. Blasse G (1980) Structure and Bonding 42: 1
49. van der Poel WAJA, Noort M, Herbich J, Coremans CJM, van der Waals JH (1984) Chem Phys Letters 103: 245
50. Barendswaard W, van Tol J, van der Waals JH (1985) Chem Phys Letters 121: 361
51. Barendswaard W, van der Walls JH (1986) Mol Phys 59: 337

52. Barendswaard W, van Tol J, Weber RT, van der Waals JH (1989) Mol Phys 67: 651
53. Coremans CJM, Groenen EJJ, van der Waals JH (1990) J Chem Phys 93: 3101
54. Freiberg A, Rebane LA (1979) J Luminescence 18/19: 702
55. Anpo M, Sunamoto M, Che M (1989) J Phys Chem 93: 1187
56. Anpo M, Tanahashi I, Kubokawa Y (1980) J Phys Chem 84: 3440; 86 (1982) 1
57. Patterson HH, Cheng J, Despres S, Sunamoto M, Anpo M (1991) J Phys Chem 95: 8813
58. Hazenkamp MF, Blasse G (1992) J Phys Chem 96: 3442
59. Hazenkamp MF, Strijbosch AWPM, Blasse G (1992) J Solid State Chem 97: 115
60. Hazenkamp MF, van Duijneveldt FB, Blasse G (1993) Chem Physics 169: 55
61. Hazenkamp MF (1992) thesis, Utrecht
62. Hazenkamp MF, Blasse G (1992) Ber Bunsenges Phys Chem 96: 1471
63. Blasse G, Dirksen GJ, Pausewang GJ, Schmidt R (1990) J Solid State Chem 88: 586
64. See, for example, Bleijenberg KC (1980) Structure and Bonding 42: 97
65. Denning RG, Norris JOW, Laing PJ (1985) Mol Phys 54: 713
66. Barker TJ, Denning RG, Thorne JRG (1987) Inorg Chem 26: 1721
67. Denning RG, Morrison ID (1991) Chem Phys Letters 180: 101
68. Denning RG (1992) Structure and Bonding 79: 215

The Doublet States in Chromium(III) Complexes. A Shell-Theoretic View

Arnout Ceulemans

Departement Scheikunde, Katholieke Universiteit Leuven, Celestijnenlaan 200F, B-3001 Leuven, Belgium

Table of Contents

Topics in Current Chemistry, Vol. 171

In this review the electronic origins belonging to the $(t_{2g})^3$ configuration in hexacoordinated Cr(III) complexes are described from the perspective of a shell-theoretic approach. The states of the shell are characterized by seven quantum numbers. Five of these relate to the electron spin, pseudo-angular momentum and spatial symmetry. In addition there are two quasi-spin quantum numbers, specifying the configurational symmetry of half-filled shell states. The review examines the spectroscopic impact of this rich quantum structure. Selection rules translate the internal symmetries of the shell into observable splitting patterns. Effects of interelectronic repulsion, spin-orbit coupling, ligand fields and external magnetic fields are discussed.

1 Introduction

The $(t_{2g})^3$ configuration in hexacoordinated complexes of Cr(III) and other d^3 transition-metal ions gives rise to a ${}^4A_{2g}$ ground state and three low-lying doublet states which are labeled as ${}^2E_g$, ${}^2T_{1g}$ and ${}^2T_{2g}$. These multiplets constitute a remarkable electronic system which has fascinated the practitioners of ligand field theory ever since the pioneering studies of Sugano and Tanabe, Schläfer and Gliemann and many others [1–4]. The classical apparatus of ligand field theory provides the tools to describe the spectral and magnetic features of this system. Term diagrams such as the one shown in Fig. 1 and the corresponding wavefunctions in determinantal form are familiar to generations of inorganic chemists interested in the electronic structure of transition-metal compounds.

In the present contribution we will make an attempt to relate these descriptions to a more recent view of electronic shells based on the technique of Second Quantization. We are aware of the fact that modern shell theory has not yet found its way into inorganic textbooks, partly because most treatises on this subject focus on the mathematical intricacies of the method, rather than on its conceptual basis. One of the advantages of the shell approach is that it debouches in a natural way into the concept of quasi-spin, which is an important characteristic of the states under consideration. The resulting description of the $(t_{2g})^3$ multiplets is based on not less than seven quantum numbers, specifying quasi-spin, spin, pseudo-angular momentum, and spatial symmetry. As can be expected such a structure gives rise to a rich pattern of selection rules which will determine the events recorded by the spectrometer. In this way the quantum structure of the shell makes its presence felt.

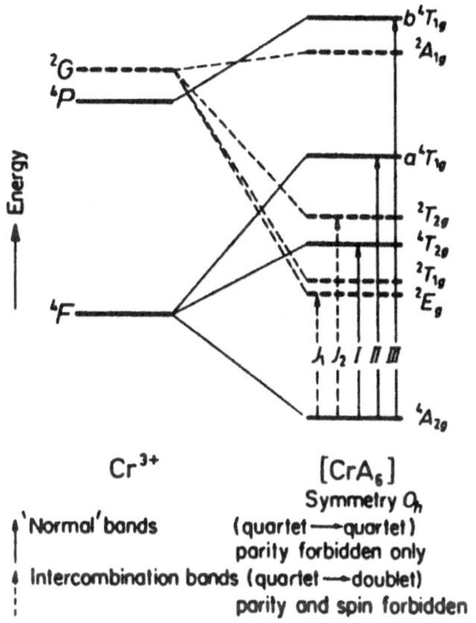

Fig. 1. Term system for Cr^{3+} in a ligand field of O_h symmetry, as depicted by Schläfer and Gliemann (^2G is the only one of the doublet terms shown). Reprinted with permission from Ref. [1], p 44, Fig. A32

The purpose of the present paper is twofold: first, to place the existing spectroscopic and magnetic data in the perspective of the theoretical shell description, second, to draw attention to some unresolved problems which require further experimental studies. Attention will be limited to the interactions which affect the electronic origins of the intercombination bands in hexacoordinated chromium(III) complexes. The wealth of information contained in the vibrational fine structure will not be discussed. A full covering of the spectroscopic material is not intended. Further data can indeed be found in existing reviews [5, 6]. It should be noted that new interesting information is also discussed in other contributions to the present volume [7].

2 Shell-Theoretic Description of the $(t_{2g})^3$ Multiplets

In this section, we will develop a theoretical description of the $(t_{2g})^3$ states using Second Quantization. The treatment is based on Judd's invaluable G.H. Dieke Memorial Lectures [8].

2.1 Short Introduction to the Formalism

The starting point is a shell consisting of a collection of spin-orbitals. Occupation of these orbitals is represented by a normalized Slater determinant. An alternative or 'second' point of view is to consider this determinant as the result of a sequence of operators acting on the unoccupied shell. Hence one defines:

$$a_\alpha^\dagger a_\beta^\dagger \cdots a_\nu^\dagger|0\rangle \equiv |\varphi_\alpha \varphi_\beta \cdots \varphi_\nu| \tag{1}$$

Here the sequence of spin-orbitals between vertical bars represents a normalized Slater determinant. The $a^\dagger$ operators are so-called creation operators and $|0\rangle$ is the vacuum state. We may undo the action of the creation operators by introducing so-called annihilation operators, as exemplified below:

$$a_\nu \cdots a_\beta a_\alpha a_\alpha^\dagger a_\beta^\dagger \cdots a_\nu^\dagger|0\rangle = |0\rangle \tag{2}$$

This implies that the bra expression $\langle 0|a_\nu \cdots a_\alpha$ is the adjoint of the corresponding ket expression $a_\alpha^\dagger \cdots a_\nu^\dagger|0\rangle$. The equivalence of the operator and determinantal formalism ensures that all operators act in accordance with the Pauli exclusion principle for fermions. The following anticommutation rules thus follow at once:

$$\begin{aligned} &a_\xi^\dagger a_\eta^\dagger + a_\eta^\dagger a_\xi^\dagger = 0 \\ &a_\xi a_\eta + a_\eta a_\xi = 0 \\ &a_\xi^\dagger a_\eta + a_\eta a_\xi^\dagger = \delta(\xi, \eta) \end{aligned} \tag{3}$$

In the final line, $\delta(\xi, \eta)$ represents the Kronecker delta. Its value is one for $\xi = \eta$, otherwise it is zero. Some further useful relations are given below:

$$\begin{aligned} &a_\xi^\dagger a_\xi^\dagger = a_\xi a_\xi = 0 \\ &a_\xi|0\rangle = 0 \end{aligned} \tag{4}$$

The operator $\sum_\xi a_\xi^\dagger a_\xi$ where the sum runs over all elements of the shell, is called the number operator. When acting on a shell state it yields the corresponding occupation number, e.g.:

$$\left(\sum_\xi a_\xi^\dagger a_\xi\right) a_\alpha^\dagger a_\beta^\dagger a_\gamma^\dagger|0\rangle = 3a_\alpha^\dagger a_\beta^\dagger a_\gamma^\dagger|0\rangle \tag{5}$$

Finally we may consider a one-particle operator $F = \sum_i f_i$. Its interaction with the shell is represented by a matrix of one-electron elements $\langle\varphi_\xi|f|\varphi_\eta\rangle$. The appropriate second-quantized form of this operator reads:

$$F \equiv \sum_{\xi,\eta} a_\xi^\dagger \langle\varphi_\xi|f|\varphi_\eta\rangle a_\eta \tag{6}$$

It can indeed be shown that this form gives rise to the Slater-Condon rules for matrix elements between multi-electron states. Further expressions for two-particle operators may be found in the literature [8,9].

2.2 Pseudo-Angular Momentum States

The familiar set of the three t_{2g} orbitals in an octahedral complex constitutes a three-dimensional shell. Classical ligand field theory has drawn attention to the fact that the matrix representation of the angular momentum operator ℓ in a p-orbital basis is equal to the matrix of $-\ell$ in the basis of the three d-orbitals with t_{2g} symmetry [2, 3]. This correspondence implies that, under a d-only assumption, t_{2g} electrons can be treated as pseudo-p electrons, yielding an interesting isomorphism between $(t_{2g})^n$ states and atomic $(p)^n$ multiplets. We will discuss this relationship later on in more detail.

At the outset though, it must be pointed out very clearly that the development of a pseudo-angular momentum theory of the t_{2g} shell does not really require a d-only assumption. Indeed the mere fact that the t_{2g} orbitals form a three-dimensional vector space implies the existence of a group of orthogonal transformations of the orbital basis and hence of pseudo-angular momentum operators. In constructing this vector space due attention must be paid to the finite point group symmetry adaptation of the final coupled quantities because this provides the link between the internal symmetry of the shell and the actual symmetry of the molecule. The parity is not relevant here since only even functions are involved. This leaves the group O of proper rotations of the octahedron.

Rotations in a vector space of three orbitals are described by the group SO(3) of orthogonal 3×3 matrices with determinant $+1$. To embed the octahedral rotation group in this covering group one needs a matrix representation of O which also consists of orthogonal and unimodular 3×3 matrices. Such a matrix representation is sometimes called the *fundamental vector representation* of the point group. In the case of O the fundamental vector representation is T_1 and not T_2. Indeed the T_1 matrices are unimodular, i.e. have determinant $+1$, while the determinants of the T_2 matrices are equal to the characters of the one-dimensional representation A_2.

$$\det(D^{T_2}(R)) = \chi^{A_2}(R) \tag{7}$$

with $R \in O$. Hence the T_2 determinants are $+1$ for the symmetry operations E, $8C_3$, $3C_2$ and -1 for $6C_2'$ and $6C_4$. A direct embedding of the irreducible T_2 representation in a covering group of unimodular matrices is therefore not possible. Fortunately however T_1 and T_2 are *associated* representations. This means that a T_2 basis can always be turned into a T_1 basis by multiplication with a pseudo-scalar of A_2 symmetry:

$$T_1 = A_2 \times T_2 \tag{8}$$

The development of a pseudo-angular momentum theory of $(t_{2g})^n$ states, which is rooted in the point group under consideration can thus proceed in the following way: one first converts the T_2 basis into a T_1 basis. The T_1 functions can then be coupled to multiplet states using the coupling theory of the full rotation group. The resulting vectors can always be adapted to the point group symmetry by inverting the direct product in Eq. 8.

For real components the coupling table corresponding to the $A_2 \times T_2$ product has the following form [3]:

$$|t_1 x\rangle = A_2 |t_2 yz\rangle$$
$$|t_1 y\rangle = A_2 |t_2 xz\rangle$$
$$|t_1 z\rangle = A_2 |t_2 xy\rangle \tag{9}$$

Here A_2 symbolizes a pseudo-scalar of A_2 symmetry, normalized to unity. The actual form of this pseudoscalar need not bother us. The only property we will have to use later on is that even powers of A_2 are equal to $+1$. Now we can proceed by defining rotation generators $\ell'_x, \ell'_y, \ell'_z$ in the standard way, as indicated in Table 1 [10]. Note that primed symbols are used here to distinguish the pseudo-operators from their true counterparts in real coordinate space. Evidently the action of the true angular momentum operators ℓ_x, ℓ_y, ℓ_z on the basis functions is ill defined since these functions contain small ligand terms.

By standard procedures the t_1 states may be diagonalized with respect to the ℓ'_z operators, yielding $|l' m_{l'}\rangle$ kets (with $l' = 1$). If one adopts a Condon and Shortley phase convention [9] these eigenkets read:

$$|t_1 0\rangle = |t_1 z\rangle$$
$$|t_1 +1\rangle = (-|t_1 x\rangle - i|t_1 y\rangle)/\sqrt{2}$$
$$|t_1 -1\rangle = (|t_1 x\rangle - i|t_1 y\rangle)/\sqrt{2} \tag{10}$$

This basis may be converted to the Fano-Racah convention [11] by multiplying all kets in the right-hand side of Eq. 10 by the same phase factor i. Adding spin results in a shell consisting of six possible states. These states may be created by operators $a^\dagger_{m_s m_{l'}}$.

The full set of $a^\dagger$ operators forms a double tensor of rank $s = 1/2$ in spin space and rank $l' = 1$ in orbital space. The $a^\dagger_{m_s m_{l'}}$ operator directly coincides with the $m_s m_{l'}$ component of this $a^\dagger$ tensor. The corresponding annihilation operators also form a double tensor of ranks $s = 1/2$ and $l' = 1$. In this case the $m_s m_{l'}$ tensor component, commonly denoted as $\tilde{a}_{m_s m_{l'}}$, is related to the $a_{-m_s -m_{l'}}$ operator in the following way:

$$\tilde{a}_{m_s m_{l'}} = (-1)^{s + l' - m_s - m_l} a_{-m_s - m_{l'}} \tag{11}$$

Table 1. Effects of pseudo-angular momentum operator on real *t*-orbitals

	$\|t_1 x\rangle$	$\|t_1 y\rangle$	$\|t_1 z\rangle$
ℓ'_x	0	$i\|t_1 z\rangle$	$-i\|t_1 y\rangle$
ℓ'_y	$-i\|t_1 z\rangle$	0	$i\|t_1 x\rangle$
ℓ'_z	$i\|t_1 y\rangle$	$-i\|t_1 x\rangle$	0

This relationship expresses the adjoint character of the annihilation operators. The effect of the time reversal operator on these tensors in the Fano-Racah phase convention is given by:

$$Ta^{\dagger}_{m_s m_{l'}} T^{-1} = (-1)^{s+l'-m_s-m_{l'}} a^{\dagger}_{-m_s-m_{l'}}$$
$$T\tilde{a}_{m_s m_{l'}} T^{-1} = (-1)^{s+l'-m_s-m_{l'}} \tilde{a}_{-m_s-m_{l'}} \tag{12}$$

Here T symbolizes Wigner's time reversal operator.

Knowing the symmetries of the operators we can now construct coupled quantities with well defined spin and orbital characteristics. As an example two-particle states with total quantum numbers S and L' can be formed as follows:

$$(a^{\dagger}a^{\dagger})^{SL'}_{M_S M_{L'}} = (-1)^{-M_S-M_{L'}}[S]^{1/2}[L']^{1/2}$$
$$\times \sum_{\xi,\eta} \begin{pmatrix} 1/2 & 1/2 & S \\ m_{s\xi} & m_{s\eta} & -M_S \end{pmatrix} \begin{pmatrix} 1 & 1 & L' \\ m_{l'_\xi} & m_{l'_\eta} & -M_{L'} \end{pmatrix} a^{\dagger}_{m_{s_\xi} m_{l'_\xi}} a^{\dagger}_{m_{s_\eta} m_{l'_\eta}} \tag{13}$$

Square brackets denote multiplicities, hence $[S] = 2S + 1$, $[L'] = 2L' + 1$. In this equation, one recognizes the Wigner $3j$ symbols which describe the coupling of the individual spin and orbital momenta. These coefficients were devised to couple unrelated systems. Since at present we are coupling equivalent electrons the resulting expressions might need renormalization or might even vanish because of the exclusion principle. Further extension of these coupling formulae to the case of three t electrons can most easily be achieved by using Racah's coefficients of fractional parentage [12]. This has been described at length in several textbooks [9]. The resulting $(t)^3$ operators in $SL'M_SM_{L'}$ quantized form are given in Table 2.

The M_S and $M_{L'}$ quantum numbers of a given ${}^{2S+1}L'$ term can be raised or lowered by using the well known shift operators $\mathscr{L}'^{\pm}$ and $\mathscr{S}^{\pm}$. The second-quantized form of these operators is as follows:

$$\mathscr{L}'^{\pm} = \sqrt{2}(a^{\dagger}_{\alpha 0}a_{\alpha \mp 1} + a^{\dagger}_{\beta 0}a_{\beta \mp 1} + a^{\dagger}_{\alpha \pm 1}a_{\alpha 0} + a^{\dagger}_{\beta \pm 1}a_{\beta 0})$$
$$\mathscr{S}^{+} = a^{\dagger}_{\alpha+1}a_{\beta+1} + a^{\dagger}_{\alpha 0}a_{\beta 0} + a^{\dagger}_{\alpha-1}a_{\beta-1}$$
$$\mathscr{S}^{-} = a^{\dagger}_{\beta+1}a_{\alpha+1} + a^{\dagger}_{\beta 0}a_{\alpha 0} + a^{\dagger}_{\beta-1}a_{\alpha-1} \tag{14}$$

The action of these operators on the expressions in Table 2 is governed by the set of commutation rules, specified in Eq. 3. Their raising or lowering character is made particularly transparent by the second-quantized formalism. As an example the $\mathscr{S}^{+}$ operator is seen to consist of $a^{\dagger}_{\alpha m_{l'}}\, a_{\beta m_{l'}}$ parts: as these parts run through the operator sequences of the $(t)^3$ multiplets they destroy a β spin on the $l'm_{l'}$ orbital and create an α spin on the same orbital, thus raising M_S by one.

A final point concerns the finite point group symmetry adaptation of the resulting states. Since the octahedral group has been rooted in the Lie group of the pseudo-angular momentum generators the standard $SO(3) \downarrow O \downarrow D_4$ subduction relations [13] can be used to obtain the symmetry adapted ΓM_Γ combinations. The S', P', and D' states thus match resp. A_1, T_1 and $E + T_2$

Table 2. $(t)^3$ multiplet operators in $SL'M_SM_{L'}$ quantized form

M_S	$M_{L'}$	
		$^4S'$
3/2	0	$a^\dagger_{\alpha+1}a^\dagger_{\alpha 0}a^\dagger_{\alpha-1}$
1/2	0	$(a^\dagger_{\alpha+1}a^\dagger_{\alpha 0}a^\dagger_{\beta-1} + a^\dagger_{\alpha+1}a^\dagger_{\beta 0}a^\dagger_{\alpha-1} + a^\dagger_{\beta+1}a^\dagger_{\alpha 0}a^\dagger_{\alpha-1})/\sqrt{3}$
− 1/2	0	$(a^\dagger_{\alpha+1}a^\dagger_{\beta 0}a^\dagger_{\beta-1} + a^\dagger_{\beta+1}a^\dagger_{\alpha 0}a^\dagger_{\beta-1} + a^\dagger_{\beta+1}a^\dagger_{\beta 0}a^\dagger_{\alpha-1})/\sqrt{3}$
− 3/2	0	$a^\dagger_{\beta+1}a^\dagger_{\beta 0}a^\dagger_{\beta-1}$
		$^2D'$
1/2	2	$-a^\dagger_{\alpha+1}a^\dagger_{\beta+1}a^\dagger_{\alpha 0}$
1/2	1	$(a^\dagger_{\alpha+1}a^\dagger_{\alpha 0}a^\dagger_{\beta 0} - a^\dagger_{\alpha+1}a^\dagger_{\beta+1}a^\dagger_{\alpha-1})/\sqrt{2}$
1/2	0	$(a^\dagger_{\alpha+1}a^\dagger_{\alpha 0}a^\dagger_{\beta-1} + a^\dagger_{\beta+1}a^\dagger_{\alpha 0}a^\dagger_{\alpha-1} - 2a^\dagger_{\alpha+1}a^\dagger_{\beta 0}a^\dagger_{\alpha-1})/\sqrt{6}$
1/2	− 1	$(a^\dagger_{\alpha+1}a^\dagger_{\alpha-1}a^\dagger_{\beta-1} - a^\dagger_{\alpha 0}a^\dagger_{\beta 0}a^\dagger_{\alpha-1})/\sqrt{2}$
1/2	− 2	$a^\dagger_{\alpha 0}a^\dagger_{\alpha-1}a^\dagger_{\beta-1}$
− 1/2	2	$-a^\dagger_{\alpha+1}a^\dagger_{\beta+1}a^\dagger_{\beta 0}$
− 1/2	1	$(a^\dagger_{\beta+1}a^\dagger_{\alpha 0}a^\dagger_{\beta 0} - a^\dagger_{\alpha+1}a^\dagger_{\beta+1}a^\dagger_{\beta-1})/\sqrt{2}$
− 1/2	0	$(-a^\dagger_{\alpha+1}a^\dagger_{\beta 0}a^\dagger_{\beta-1} + a^\dagger_{\beta+1}a^\dagger_{\beta 0}a^\dagger_{\alpha-1} + 2a^\dagger_{\beta+1}a^\dagger_{\alpha 0}a^\dagger_{\beta-1})/\sqrt{6}$
− 1/2	− 1	$(a^\dagger_{\beta+1}a^\dagger_{\alpha-1}a^\dagger_{\beta-1} - a^\dagger_{\alpha 0}a^\dagger_{\beta 0}a^\dagger_{\beta-1})/\sqrt{2}$
− 1/2	− 2	$a^\dagger_{\beta 0}a^\dagger_{\alpha-1}a^\dagger_{\beta-1}$
		$^2P'$
1/2	1	$(a^\dagger_{\alpha+1}a^\dagger_{\beta+1}a^\dagger_{\alpha-1} + a^\dagger_{\alpha+1}a^\dagger_{\alpha 0}a^\dagger_{\beta 0})/\sqrt{2}$
1/2	0	$(a^\dagger_{\alpha+1}a^\dagger_{\alpha 0}a^\dagger_{\beta-1} - a^\dagger_{\beta+1}a^\dagger_{\alpha 0}a^\dagger_{\alpha-1})/\sqrt{2}$
1/2	− 1	$(a^\dagger_{\alpha+1}a^\dagger_{\alpha-1}a^\dagger_{\beta-1} + a^\dagger_{\alpha 0}a^\dagger_{\beta 0}a^\dagger_{\alpha-1})/\sqrt{2}$
− 1/2	1	$(a^\dagger_{\alpha+1}a^\dagger_{\beta+1}a^\dagger_{\beta-1} + a^\dagger_{\beta+1}a^\dagger_{\alpha 0}a^\dagger_{\beta 0})/\sqrt{2}$
− 1/2	0	$(a^\dagger_{\alpha+1}a^\dagger_{\beta 0}a^\dagger_{\beta-1} - a^\dagger_{\beta+1}a^\dagger_{\beta 0}a^\dagger_{\alpha-1})/\sqrt{2}$
− 1/2	− 1	$(a^\dagger_{\beta+1}a^\dagger_{\alpha-1}a^\dagger_{\beta-1} + a^\dagger_{\alpha 0}a^\dagger_{\beta 0}a^\dagger_{\beta-1})/\sqrt{2}$

representations. The subduction rules for real components are independent of spin and read as follows:

$$|S'A_1a_1\rangle = |S'0\rangle$$
$$|P'T_1x\rangle = (|P' - 1\rangle - |P' + 1\rangle)/\sqrt{2}$$
$$|P'T_1y\rangle = (i|P' + 1\rangle + i|P' - 1\rangle)/\sqrt{2}$$
$$|P'T_1z\rangle = |P'0\rangle$$
$$|D'E\theta\rangle = |D'0\rangle$$
$$|D'E\varepsilon\rangle = (|D' + 2\rangle + |D' - 2\rangle)/\sqrt{2}$$
$$|D'T_2yz\rangle = (i|D' + 1\rangle + i|D' - 1\rangle)/\sqrt{2}$$
$$|D'T_2xz\rangle = (-|D' + 1\rangle + |D' - 1\rangle)/\sqrt{2}$$
$$|D'T_2xy\rangle = (-i|D' + 2\rangle + i|D' - 2\rangle)/\sqrt{2} \qquad (15)$$

The ket functions in these expressions are linear combinations of triple products of the t_{2g} orbitals times the third power of the pseudo-scalar A_2. To eliminate this factor we multiply both sides of the equations in (15) by A_2. The even power of A_2 in the right-hand side of the equations is equal to $+1$. On the left-hand side this leads to a Kronecker product $A_2 \times \Gamma$ which can be worked out according to the appropriate coupling tables. Following Griffith [3] one has:

$$\begin{aligned}
&A_2|S'A_1a_1\rangle = |S'A_2a_2\rangle\\
&A_2|P'T_1x\rangle = |P'T_2yz\rangle\\
&A_2|P'T_1y\rangle = |P'T_2xz\rangle\\
&A_2|P'T_1z\rangle = |P'T_2xy\rangle\\
&A_2|D'E\theta\rangle = -|D'E\varepsilon\rangle\\
&A_2|D'E\varepsilon\rangle = |D'E\theta\rangle\\
&A_2|D'T_2yz\rangle = |D'T_1x\rangle\\
&A_2|D'T_2xz\rangle = |D'T_1y\rangle\\
&A_2|D'T_2xy\rangle = |D'T_1z\rangle
\end{aligned} \tag{16}$$

Combination of Eqs. 15 and 16 with the expressions in Table 2 thus finally yields the symmetry adapted $(t_{2g})^3$ multiplets ${}^4A_{2g}$, ${}^2E_g + {}^2T_{1g}$, ${}^2T_{2g}$ described by five quantum numbers $|SL'\Gamma M_S M_\Gamma\rangle$. These resulting states are identical – within a multiplet dependent phase factor – to the state functions published by Griffith [3].

2.3 Quasi-Spin States

Further exploration of the internal symmetries of the $(t_{2g})^3$ multiplets introduces us to the concept of quasi-spin. For a given l' shell one defines an operator $\mathcal{Q}$ with components $\mathcal{Q}^+$, $\mathcal{Q}^-$, $\mathcal{Q}_z$ as given in Eq. 17.

$$\begin{aligned}
&\mathcal{Q}^+ = 1/2[s]^{1/2}[l']^{1/2}(a^\dagger a^\dagger)^{00}\\
&\mathcal{Q}^- = -1/2[s]^{1/2}[l']^{1/2}(aa)^{00}\\
&\mathcal{Q}_z = -1/4[s]^{1/2}[l']^{1/2}\{(a^\dagger a)^{00} + (aa^\dagger)^{00}\}
\end{aligned} \tag{17}$$

The 00 superscripts indicate that the $a^\dagger$ and a operators are coupled to quantities of rank zero in spin and orbit space. The scalar products in this equation can be evaluated by means of the general expansion formula in Eq. 13. For the case of an $l' = 1$ shell the following expressions are obtained:

$$\begin{aligned}
\mathcal{Q}^+ &= a^\dagger_{\alpha+1}a^\dagger_{\beta-1} - a^\dagger_{\alpha 0}a^\dagger_{\beta 0} + a^\dagger_{\alpha-1}a^\dagger_{\beta+1}\\
\mathcal{Q}^- &= a_{\alpha+1}a_{\beta-1} - a_{\alpha 0}a_{\beta 0} + a_{\alpha-1}a_{\beta+1}\\
2\mathcal{Q}_z &= a^\dagger_{\alpha+1}a_{\alpha+1} + a^\dagger_{\beta+1}a_{\beta+1} + a^\dagger_{\alpha 0}a_{\alpha 0} + a^\dagger_{\beta 0}a_{\beta 0}\\
&\quad + a^\dagger_{\alpha-1}a_{\alpha-1} + a^\dagger_{\beta-1}a_{\beta-1} - 3
\end{aligned} \tag{18}$$

These operators obey commutation rules which are identical to the commutation rules for angular momentum or spin operators, hence the name quasi-spin

$$[\mathcal{Q}^+, \mathcal{Q}^-] = 2\mathcal{Q}_z$$
$$[\mathcal{Q}_z, \mathcal{Q}^+] = \mathcal{Q}^+$$
$$[\mathcal{Q}^-, \mathcal{Q}_z] = \mathcal{Q}^- \tag{19}$$

The scalar product of the resultant quasi-spin vector $\mathcal{Q}$ is given by:

$$\mathcal{Q} \cdot \mathcal{Q} = \mathcal{Q}_z^2 - \mathcal{Q}_z + \mathcal{Q}^+ \mathcal{Q}^- \tag{20}$$

The $\mathcal{Q}^+$ and $\mathcal{Q}^-$ operators are seen to create or destroy pairs of particles coupled to zero spin and orbital momentum. Hence these ladder operators move across the configurations changing particle numbers by two while keeping S and L' assignments fixed. In contrast the $\mathcal{Q}_z$ operator which determines the M_Q quantum characteristic, leaves the configuration unchanged. It is seen to contain the number operator, defined earlier in Eq. 5, and thus measures the occupation number of a given configuration. One has in general:

$$\mathcal{Q}_z |(l')^n\rangle = 1/2(n - 2l' - 1)|(l')^n\rangle \tag{21}$$

Hence for a half-filled shell M_Q equals zero. In the case of the t_{2g} shell one manifold of quasi-spin states will comprise the $(t)^1$, $(t)^3$ and $(t)^5$ configurations while the other contains the even-electron $(t)^0$, $(t)^2$, $(t)^4$ and $(t)^6$ configurations. A simple bookkeeping of M_Q, S and L' labels then leads to the identification of one quasi-quartet, one quasi-triplet, two quasi-doublets and two quasi-singlets. These results are displayed in Table 3. All these assignments may of course be verified at once by operating with $\mathcal{Q} \cdot \mathcal{Q}$ on the multiplets yielding the associated $Q(Q + 1)$ eigenvalues.

Focusing now our attention to our set of $(t_{2g})^3$ states we observe that $^2D'$ and $^4S'$ are the $M_Q = 0$ components of a quasi-spin singlet while $^2P'$ is the $M_Q = 0$ component of a quasi-spin triplet. The full quantum structure of the $(t_{2g})^3$ multiplets thus involves seven labels $|Q M_Q S L' \Gamma M_S M_\Gamma\rangle$, albeit the M_Q label is redundant since all states share the same M_Q value.

Table 3. Quasi-spin manifolds of the t-shell

		Q					
$(t)^n$	M_Q	3/2	1	1/2	1/2	0	0
0	−3/2	$^1S'$					
1	−1		$^2P'$				
2	−1/2	$^1S'$		$^1D'$	$^3P'$		
3	0		$^2P'$			$^2D'$	$^4S'$
4	+1/2	$^1S'$		$^1D'$	$^3P'$		
5	+1		$^2P'$				
6	+3/2	$^1S'$					

The quasi-spin label is perhaps the most exotic quantum characteristic of these states. It is in any case a true shell characteristic since it extends over several configurations. Most importantly it gives rise to very strong selection rules as we will demonstrate in the next section.

2.4 Quasi-Spin Selection Rules

The $(t_{2g})^3$ multiplets are exposed to a variety of interactions ranging from interelectronic repulsion, spin-orbit coupling, and ligand field effects to vibronic coupling, Zeeman splittings, electromagnetic radiation, and last but not least configuration interactions. In principle each of the previously defined quantum labels can give rise to selection rules. These rules will be effective to the extent that the interactions themselves respond in a proper way to the corresponding symmetry operations. We will discuss the various interaction mechanisms in detail in the subsequent sections, devoting at present special attention to the strongest and most unusual selection rule of all, i.e. the quasi-spin selection rule.

The discussion of this rule is facilitated by introducing the concept of *triple tensor* [8]. Equisymmetric creation and annihilation operators $a^{\dagger}_{m_s m_{l'}}$ and $\tilde{a}_{m_s m_{l'}}$ can indeed be shown to be the $m_q = +1/2$ and $m_q = -1/2$ partners, respectively, of a quasi-spin doublet ($q = 1/2$). As such these components may be grouped together in a covering triple tensor $(a)^{qsl'}$ containing $[q][s][l']$ elements. One-electron operators, as indicated in Eq. 6, define a tensor product of these $(a)^{qsl'}$ quantities. Since this product is a bilinear form of creation and annihilation operators its total M_Q value must be zero. The task to identify the overall Q value is much simplified by the fact that the quasi-spin operators are totally scalar with respect to spin and space properties. As a result the coupling of creation and annihilation operators will be a quasi-spin singlet or a quasi-spin triplet if the product form is antisymmetric or symmetric under exchange of associated $m_q = +1/2$ and $m_q = -1/2$ partners.

To examine this exchange behavior for an arbitrary one-electron operator we start by rewriting the general expression of Eq. 6 in a sum over symplectic pairs. In the following ξ and η are assumed to be compound indices for resp. $(m_{s_\xi}, m_{l_\xi'})$ and $(m_{s_\eta}, m_{l_\eta'})$. Also $-\xi$ refers to $(-m_{s_\xi}, -m_{l_\xi'})$. The matrix element $\langle \varphi_\xi | f | \varphi_\eta \rangle$ is abbreviated as $F_{\xi\eta}$.

$$\begin{aligned} F &= \sum_{\xi,\eta} F_{\xi\eta} a^{\dagger}_{\xi} a_{\eta} \\ &= 1/2 \sum_{\xi,\eta} (F_{\xi\eta} a^{\dagger}_{\xi} a_{\eta} + F_{-\eta-\xi} a^{\dagger}_{-\eta} a_{-\xi}) \\ &= 1/2 \sum_{\xi,\eta} (F_{\xi\eta} a^{\dagger}_{\xi} a_{\eta} - F_{-\eta-\xi} a_{-\xi} a^{\dagger}_{-\eta}) + 1/2 \sum_{\xi} F_{\xi\xi} \end{aligned} \quad (22)$$

The annihilation operators in this equation must now be represented in their proper tensorial form as defined in Eq. 11. This yields:

$$F - 1/2\sum_{\xi} F_{\xi\xi} = 1/2\sum_{\xi,\eta}(-1)^{-s-l'}$$

$$\times\left\{(-1)^{-m_{s_\eta}-m_{l'_\eta}} F_{\xi\eta} a_\xi^\dagger \tilde{a}_{-\eta} - (-1)^{m_{s_\xi}+m_{l'_\xi}} F_{-\eta-\xi} \tilde{a}_\xi a_{-\eta}^\dagger\right\}$$

(23)

The hermitean character of the F operator allows the replacement of $F_{-\eta-\xi}$ by the complex conjugate of $F_{-\xi-\eta}$. This matrix element can be related to $F_{\xi\eta}$ if it is assumed that F either commutes or anticommutes with time reversal.

$$F_{-\eta-\xi} = \bar{F}_{-\xi-\eta} = (-1)^{m_{s_\eta}+m_{l'_\eta}-m_{s_\xi}-m_{l'_\xi}+x} F_{\xi\eta} \tag{24}$$

where $(-1)^x$ equals $+1$ if F is symmetric under time reversal and -1 if F is antisymmetric under time reversal. The remainder of the phase factor in Eq. 24 stems from the time reversal behavior of the $\langle\varphi_{-\xi}|$ and $|\varphi_{-\eta}\rangle$ parts of the bracket and follows the conventions in Eq. 12. Substituting this result in Eq. 23 finally yields:

$$F - 1/2\sum_{\xi} F_{\xi\xi} = 1/2\sum_{\xi,\eta}(-1)^{-s-l'-m_{s_\eta}-m_{l'_\eta}} F_{\xi\eta}(a_\xi^\dagger \tilde{a}_{-\eta} + (-1)^x \tilde{a}_\xi a_{-\eta}^\dagger) \tag{25}$$

In this expression the F operator is broken up into its irreducible quasi-spin parts. The behavior of F under time reversal thereby proves to be the determining factor. Two cases are possible.

1) F and T commute. Equation 25 then becomes:

$$F(\text{time-even}) = 1/2\sum_{\xi} F_{\xi\xi} + 1/2\sum_{\xi,\eta}(-1)^{-s-l'-m_{s_\eta}-m_{l'_\eta}} F_{\xi\eta}$$

$$\times(a_\xi^\dagger \tilde{a}_{-\eta} + \tilde{a}_\xi a_{-\eta}^\dagger) \tag{26}$$

The first term in this expression is proportional to the trace of the F matrix. This is a scalar and therefore has the three quantum numbers Q, S, L' equal to zero. The second term is clearly symmetric under exchange of the associated m_q partners in the operator part. It thus corresponds to a quasi-spin triplet.

2) F and T anticommute. In this case the trace of the F matrix must vanish because the time-reversed elements $F_{\xi\xi}$ and $F_{-\xi-\xi}$ will cancel. Equation 25 then yields:

$$F(\text{time-odd}) = 1/2\sum_{\xi,\eta}(-1)^{-s-l'-m_{s_\eta}-m_{l'_\eta}} F_{\xi\eta}(a_\xi^\dagger \tilde{a}_{-\eta} - \tilde{a}_\xi a_{-\eta}^\dagger) \tag{27}$$

Clearly this sum is made up of antisymmetric terms only. It thus corresponds to an irreducible quasi-spin singlet.

Knowing the quasi-spin properties of F we can now turn our attention to the associated selection rules. According to the Wigner-Eckart theorem acting in quasi-spin space the selection of an interaction element of an operator $|K\,0|$

between multiplet terms $|QM_Q\rangle$ and $|Q'M_{Q'}\rangle$ is determined by the $3j$ symbol [14]:

$$\begin{pmatrix} Q & K & Q' \\ -M_Q & 0 & M_{Q'} \end{pmatrix}$$

This symbol imposes a triangular condition on Q, K, Q'. In the case of half-filled shell states all M_Q values must be zero. The triangular condition then becomes more stringent because of the additional requirement that the sum $Q + K + Q'$ be even. This implies:

$$(l')^{2l'+1}\colon\; K = 0 \rightarrow \Delta Q = 0$$

$$K = 1 \rightarrow \Delta Q = \pm 1 \tag{28}$$

The results for one-electron operators acting in half-filled shell states thus may be summarized as follows:

1) Interaction elements between half-filled shell states with different quasi-spin character will be zero for time-odd one-electron operators.
2) Off-diagonal elements between half-filled shell states with identical quasi-spin character will be zero for time-even one-electron operators.
3) Diagonal interaction elements will be zero for time-even one-electron operators that are non-totally symmetric scalars in spin and orbit space.

The truly remarkable feature of these selection rules is that they only depend on the time-reversal and scalar character of the one-electron operator. This results in very strict quantum conditions which are of paramount importance for spectroscopy and magnetism of the $(t_{2g})^3$ systems – as we intend to show in the next sections.

We note that similar selection rules have been derived on the basis of determinantal product states, using the expansion theorem of Laplace [15]. The relationship between both formalisms is still under study [16].

Finally it should also be mentioned that the quasi-spin treatment can be extended to the two-particle Coulomb interaction. This will not be considered here since in the case of the $(t_{2g})^3$ multiplets the quasi-spin characteristics of the Coulomb operator do not give rise to additional selection rules.

3 Interelectronic Repulsion

In this section, we will examine the role of interelectronic repulsion in the perspective of the internal symmetries of the shell. The key observation is that in a d-only approximation – i.e. if the t_{2g}-orbital functions can be written as products of a common radial part and a spherical harmonic angular function of rank two – the interelectronic repulsion operator and the pseudo-angular momentum operators commute [2]. This implies that the dominant part of the repulsion will be controlled by stringent pseudo-spherical selection rules. More importantly yet it also means that small subsidiary effects such as anisotropic or multi-centre contributions are directly and separately observable via the splitting of pseudo-spherical degeneracies. Two examples of such term splittings which will appear to be sensitive probes for specific repulsion interactions will be discussed.

3.1 The Octahedral Splitting of the $^2D'$ Term

In the d-only approximation the two crystal field levels of the $^2D'$ manifold, ^2E_g and $^2T_{1g}$, retain their pseudo-spherical degeneracy as a result of the commutation of the interelectronic repulsion and pseudo-angular momentum operators. Experimentally though a substantial energy splitting ($\sim 600\ \text{cm}^{-1}$) is clearly observed. The results collected in Table 4 show that this splitting is nearly constant over a wide range of 10 Dq values [17–22]. The $[CrF_6]^{3-}$ complex is exceptional in that its $^2D'$ splitting is much larger ($1600\ \text{cm}^{-1}$).

The subsequent discussion will focus on two different repulsion mechanisms which may lift the pseudo-spherical degeneracy of the $^2D'$ term: ligand delocalization and configuration interaction.

The *ligand delocalization* mechanism allows the t_{2g} functions to deviate from the d-only limit by mixing in ligand π-orbitals. Three independent parameters, the so-called a, b, j strong-field parameters of Griffith [3], are needed to describe interelectronic repulsion in the delocalized t_{2g} shell. The following expressions are obtained:

$$
\begin{aligned}
&{}^4S'\colon\ E(^4A_{2g}) = 3b - 3j \\
&{}^2D'\colon\ \begin{cases} E(^2E_g) = 3b \\ E(^2T_{1g}) = a + 2b - 2j \end{cases} \\
&{}^2P'\colon\ E(^2T_{2g}) = a + 2b
\end{aligned}
\tag{29}
$$

Clearly in this scheme the degeneracy of ^2E_g and $^2T_{1g}$ is not maintained, unless $a - b = 2j$ which is true for the d-only limit. SCF calculations show [23, 24] that

Table 4. Splitting[a] of the $^2D'$ term in octahedral Cr(III) complexes

	10 Dq	$E(^2T_{1g}) - E(^2E_g)$	lattice	Ref.
$[CrCl_6]^{3-}$	11865	629	Cs_2NaYCl_6	[17]
$[CrF_6]^{3-}$	15600	1602	K_2NaGaF_6	[18]
$[Cr(NCS)_6]^{3-}$	17700	724	$K_3[Cr(NCS)_6]$	[19]
$[Cr(NH_3)_6]^{3+}$	21500	591	$[Cr(NH_3)_6]CdCl_5$	[20]
$[Cr(en)_3]^{3+}$	22300	567	$2[Ir(en)_3]Cl_3 \cdot KCl \cdot 6H_2O$	[21]
$[Cr(CN)_6]^{3-}$	26600	625	$K_3[Cr(CN)_6]$	[22]

[a] $E(^2T_{1g}) - E(^2E_g)$ represents the difference between the averages of the electronic origins for the $^2T_{1g} \leftarrow {}^4A_{2g}$ and $^2E_g \leftarrow {}^4A_{2g}$ transitions (in cm^{-1}).

Table 5. Ab initio calculations of interelectronic repulsion[a] in the t_{2g}-shell

	a	b	j	$a-b-2j$ (cm^{-1})	% Cr3d
$[CrF_6]^{3-}$	0.841660	0.775509	0.033116	−18	98
$[Cr(CN)_6]^{3-}$	0.790337	0.729320	0.030545	−16	99

[a] The columns a, b, j denote the strong field Griffith parameters for the t_{2g}-shell (values in Hartree). % Cr3d refers to the normalized total-gross population of the t_{2g} orbitals for the average d^3 configuration. Data taken from [23].

the extent of the delocalization is extremely small, the metal-d contribution to the t_{2g} orbitals usually being close to 100% (see Table 5). Corresponding term splittings are calculated to be very small (10 to 20 cm^{-1}) with $^2T_{1g}$ always below ^2E_g. The delocalization mechanism is thus clearly unable to account for the spectral observations in Table 4.

Before leaving the Griffith parametrization in Eq. 29 we would like to draw attention to one of its intriguing properties: the weighted average of the ^2E_g and $^2T_{1g}$ components, which is nothing more than the centre of the $^2D'$ term, is found to coincide with the average of the $(t_{2g})^3$ configuration (see Eq. 30). It is conceivable that

$$\frac{4}{20}E(^4S') + \frac{10}{20}E(^2D') + \frac{6}{20}E(^2P') = E(^2D') \tag{30}$$

a further exploration of the dynamical symmetry of the shell may provide an explanation for this remarkable coincidence.

Configuration interaction is the generally accepted mechanism to explain the octahedral splitting of the $^2D'$ term. The largest interaction element of the repulsion operator is found between the $^2E_g(t_{2g}^3)$ and $^2E_g(t_{2g}^2(^1A_{1g})e_g)$ states. This rationalizes why the ^2E_g component lies at lower energy. Semi-empirical ligand field calculations based on the complete strong-field matrices indeed yield term splittings of the right order of magnitude over a wide 10 Dq range. These results are also reproducible by extended SCF methods including configuration interaction. As an example an ab initio study [25] of the $[Cr(NH_3)_6]^{3+}$ complex ion at a limited CI level has yielded a $^2D'$ splitting of 900 cm^{-1}. In this study the strong field interaction matrices were calculated on the basis of SCF results for the average d^3 configuration and diagonalized. A more rigorous CI treatment of the $[CrF_6]^{3-}$ complex by Pierloot and Vanquickenborne [26] produced $^2D'$ splitting energies of 1000 to 1300 cm^{-1} depending on the number of correlated electrons and the nature of the active orbital space.

In spite of this considerable computational effort, calculations so far have failed to elucidate the nature of the exceptionally large splitting in the $[CrF_6]^{3-}$ complex. Inspection of the Tanabe-Sugano correlation diagram for the d^3 ions shows that the $[CrF_6]^{3-}$ complex is situated in the region where the first spin-allowed $^4T_{2g} \leftarrow {}^4A_{2g}$ transition crosses the $^2D'$ band system. In fact, the magneto-optical study of Dubicki et al. [18] assigns the origins of the

$^4T_{2g} \rightarrow {}^4A_{2g}$ transition to between the $^2T_{1g} \leftarrow {}^4A_{2g}$ and $^2E_g \leftarrow {}^4A_{2g}$ bands. Spin-orbit coupling between the $^2D'$ components and the $^4T_{2g}$ intruder level will give rise to a broad and complex absorption spectrum of the crossing region. Detailed investigations of this exceptional splitting mechanism are still lacking.

3.2 The Tetragonal Splitting of the ^2E_g State

The first coordination sphere of acido-pentamine $[Cr(NH_3)_5X]^{2+}$ complexes has C_{4v} symmetry and this leads to a tetragonal resolution of the ^2E_g state into $^2A_1$ and $^2B_1$ components. The classical ligand field model predicts that these components will be virtually degenerate. This is based on a combination of pseudo-spherical and quasi-spin selection rules of the shell and will be discussed later on in Sect. 5.2. At present we welcome this example of a pseudo-degeneracy as another opportunity to observe fine details of the interelectronic repulsion interaction which have the proper anisotropy to induce a splitting of the ^2E_g term.

In the spectroscopic investigations of Flint and Matthews [27] on a series of $[Cr(NH_3)_5X]^{2+}$ complexes a ^2E_g splitting of 100–300 cm^{-1} was reported. This is an order of magnitude larger than could be accounted for by conventional ligand field theory. Later on these data were called into question [28], until Riesen [29] finally presented unequivocal evidence for the large ^2E_g splitting using site-selective luminescence and excitation spectroscopy. The term splittings for the halo-pentamine series [30] are listed in Table 6. The splitting increases from the chloro to the iodo compound. Unfortunately the fluoro member is still lacking. Measurements on $[Cr(NH_3)_5F](ClO_4)_2$ point to the presence of a cluster of zero phonon transitions in the $^2E_g \leftarrow {}^4A_{2g}$ region which seem to arise from several nonequivalent sites in the crystal [30]. It should also be emphasized that in all cases only the absolute value of the splitting has been determined. It is taken for granted that the $^2A_1$ component lies below $^2B_1$, but this has not yet been confirmed experimentally.

Table 6. Splitting[a] of the ^2E_g term in tetragonal $[Cr(NH_3)_5X]X_2$ complexes

X	$E(^2B_1) - E(^2A_1)$
Cl	179
Br	224
I	307

[a] Energy gaps (in cm^{-1}) were taken from the work of Schmidtke et al. [30] and are in close coincidence with the earlier results of Flint and Matthews [27]. There is no experimental proof for the positive sign of the splitting.

In his original work Flint [27] suggested that the anisotropy of interelectronic repulsion in C_{4v} complexes might be responsible for the splitting. In a first-order treatment the energy difference between the two components is given by a difference of the K exchange integrals for the $e(xz, yz)$ and $b_2(xy)$ components of the t_{2g} set:

$$E(^2B_1) - E(^2A_1) = 2(K_{b_2e} - K_{ee}) \tag{31}$$

In tetragonal symmetry the equivalence of the three t_{2g} orbitals is broken so that in principle a non-zero energy difference is possible. When comparing this expression with the previous treatment of the $^2D'$ splitting using the strong-field a, b, j parameters (see Eq. 29), an essential difference is observed. In octahedral symmetry all three t_{2g} orbitals are symmetry equivalent, and thus will have the same radial function on the metal and the same extent of delocalization over the ligand. As such the metal-centered part of the $a - b - 2j$ energy difference between the two $^2D'$ sublevels should vanish, exactly as in the d-only approximation. Only the weak more-centered parts of the repulsion integrals could contribute to the splitting. In the present tetragonal example however the e and b_2 orbitals are no longer symmetry equivalent: they have different radial metal-centered functions and unequal ligand contributions. As a result the important metal-centered parts of the K_{b_2e} and K_{ee} integrals will diverge. This can give rise to a more pronounced energy gap.

Schmidtke et al. [30] presented a simplified model of the K integrals, which retains the Stevens delocalization coefficients featuring in the LCAO expansion of the metal d-orbitals over the ligands but does not allow for differences in the radial parts of the metal function. More-center and ligand-centered contributions are neglected. In this approximation the K integrals can be expressed as follows:

$$K_{b_2e} = \tau_{b_2}^2 \tau_e^2 (3B + C)$$

$$K_{ee} = \tau_e^4 (3B + C) \tag{32}$$

Here τ_{b_2} and τ_e represent the coefficients of the b_2 and e-type $3d$-functions in the appropriate LCAO expansion. B and C are the spherical Racah parameters of the d-only model. The point made by Schmidtke et al. [30] is that small deviations of the d-only limit may indeed give rise to sizeable splittings. As an example parameter values $\tau_{b_2} = 1$, $\tau_e = 0.992$ can reproduce level spacings of $200\ \mathrm{cm}^{-1}$ in agreement with experiment. The model also explains why the splitting increases from the more ionic chloro to the more covalent iodo substituent.

As a parametrization scheme this model is a typical intermediate between the crude spherical parametrization of the B and C parameters and the complete (and therefore impractical) set of the tetragonal repulsion integrals [31]. A similar strategy was used before by Koide and Pryce [32] in their treatment of octahedral d^5-complexes; they showed that quasi-degenerate terms of the half-filled d-shell configuration could be splitted by introducing a covalency difference between the t_{2g} and e_g orbitals.

While such a model offers a convenient way to reduce the number of parameters in a low-symmetry environment, it should be kept in mind that differences in the K integrals may be due not only to differences in delocalization. Other possible factors are the size of the intervening ligand orbital and the difference in the radial functions. Configuration interaction may further complicate the picture. Ab initio calculations [25] on *trans*-$[Cr(NH_3)_4Cl_2]^+$, with approximate D_{4h} symmetry, show that the e orbitals are indeed more covalent than the b_2 orbitals ($\tau_e = 0.985$, $\tau_{b_2} = 0.996$) and that this may explain the tetragonal splitting of the ^2E_g term, in accordance with the Schmidtke model [30] for $[Cr(NH_3)_5Cl]^{2+}$. However in the mono- and *trans* di-substituted fluoro-complexes the calculations point to an inverse splitting with $^2B_1$ below $^2A_1$, in spite of a similar covalency effect. This prediction clearly calls for further spectral studies of the fluoroamine complexes.

4 Spin-Orbit Coupling

Spin-orbit coupling (s.o.c.) interactions within the $(t_{2g})^3$ configuration are governed by two selection rules: a quasi-spin $\Delta Q = \pm 1$ rule and a pseudo-angular momentum $\Delta J' = 0$ rule. These rules can be used to construct a correlation diagram between weak and strong coupling limits, as we will show below. Attention will also be devoted to the splitting of spin-orbit levels in anisotropic fields.

4.1 Selection Rules

The s.o.c. operator is a one-electron operator which is even under time reversal, and non-totally symmetric in spin and orbit space. The trace of the spin-orbit coupling matrix for the t_{2g}-shell thus vanishes. As a result the s.o.c. operator is found to transform as the $M_K = 0$ component of a pure quasi-spin triplet (Cf. Eq. 26). Application of the selection rule in Eq. 28 shows that allowed matrix elements must involve a change of one unit in quasi-spin character, i.e. $\Delta Q = \pm 1$. Since $^4S'$ and $^2D'$ are both quasi-spin singlets while $^2P'$ is a quasi-spin triplet, s.o.c. interactions will be as follows:

$$\begin{aligned} \Delta Q = \pm 1 \qquad & ^4S' \not\leftrightarrow {}^2D' \\ & ^4S' \leftrightarrow {}^2P' \\ & ^2D' \leftrightarrow {}^2P' \end{aligned} \tag{33}$$

In a d-only approximation a further selection rule can be invoked, based on pseudo-angular momenta. This requires a comparison between the actions of the pseudo and true angular momentum operators in the basis of the real

Table 7. Effects of angular momentum operator on real d-orbitals (in units of $\hbar$)

	$\lvert d_{yz}\rangle$	$\lvert d_{xz}\rangle$	$\lvert d_{xy}\rangle$
ℓ_x	$i\lvert d_{x^2-y^2}\rangle + i\sqrt{3}\lvert d_{z^2}\rangle$	$-i\lvert d_{xy}\rangle$	$i\lvert d_{xz}\rangle$
ℓ_y	$i\lvert d_{xy}\rangle$	$i\lvert d_{x^2-y^2}\rangle - i\sqrt{3}\lvert d_{z^2}\rangle$	$-i\lvert d_{yz}\rangle$
ℓ_z	$-i\lvert d_{xz}\rangle$	$i\lvert d_{yz}\rangle$	$-2i\lvert d_{x^2-y^2}\rangle$

d-orbitals with t_{2g} symmetry. The effect of the pseudo-angular momentum operator was defined before in Table 1. In Table 7 we now present the effect of the true angular momentum operator [10].

Although the two tables are quite different, a remarkable coincidence is observed if one restricts attention to the matrix elements *within* the t_{2g}-shell [3]. Indeed for these matrix elements one has:

$$\langle d_i \lvert l_j \rvert d_k\rangle = -\langle t_i \lvert l'_j \rvert t_k\rangle \tag{34}$$

where i and k denote the xz, yz, and xy components of t_{2g}. This relationship implies that in a d-only approximation the pseudo-angular momentum operator can function as a genuine angular momentum operator. As a result the spin-orbit levels of the $^{2S+1}L'$ terms may be characterized by pseudo-J' values, reflecting the vector addition of S and L' momenta. In this way five spin-orbit levels are obtained, viz. $^4S'_{3/2}$, $^2D'_{5/2}$, $^2D'_{3/2}$, $^2P'_{3/2}$, $^2P'_{1/2}$. As in atomic spectroscopy s.o.c. interactions between these levels will be limited by a $\Delta J' = 0$ selection rule. Hence s.o.c. matrix elements that are allowed by the ΔQ and $\Delta J'$ selection rules are restricted to the following interactions:

$$\Delta J' = 0, \quad \Delta Q = \pm 1 \qquad {}^4S'_{3/2} \leftrightarrow {}^2P'_{3/2}$$
$$ {}^2D'_{3/2} \leftrightarrow {}^2P'_{3/2} \tag{35}$$

In the following section we will see how these rules materialize in the actual spectrum.

4.2 Weak vs Strong Spin-Orbit Coupling

Figure 2 shows a correlation diagram between limits of weak and strong spin-orbit interactions. The left hand side of the diagram corresponds to the familiar sequence of doublet levels in the absence of spin-orbit coupling. In this scheme the $^2D'$ term is resolved into its crystal field components ^2E_g and $^2T_{1g}$ due to configuration interaction with higher excited states (Cf. Sect. 3.1). All levels are further characterized as spin representations of the octahedral double group $\mathcal{O}^*$. The appropriate labels were denoted as E', E'', U' in the Griffith [3] notation. This is resp. $\Gamma_6, \Gamma_7, \Gamma_8$ in the Bethe notation. The right-hand side of the diagram shows the limit of strong spin-orbit coupling. In this limit the $^2P'$

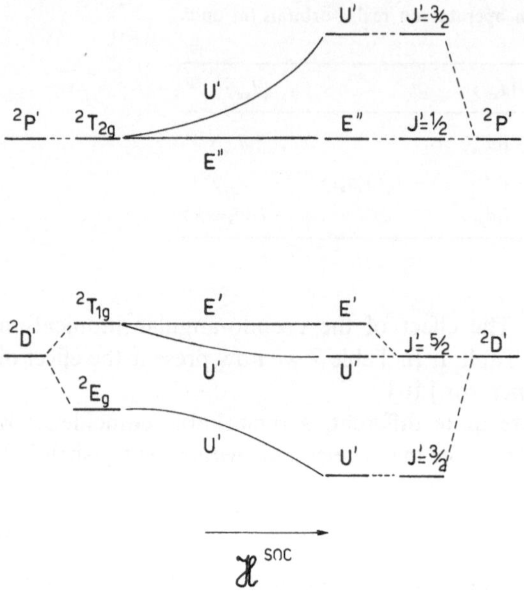

Fig. 2. Correlation diagram of the doublet states between weak (*left*) and strong (*right*) spin-orbit coupling. In the strong coupling limit the splitting pattern is determined by the pseudo-J' quantum number

and $^2D'$ states are resolved into spin-orbit components, characterized by pseudo-J' values. The zero-field splitting of these states is entirely due to second-order interactions between the $J' = 3/2$ levels. This is a direct consequence of the quasi-spin and pseudo-angular momentum selection rules of Eqs. 33 and 35. The branching rules for the J' levels in $\mathcal{O}^*$ are obtained from the standard $SU_2 \downarrow \mathcal{O}^*$ subduction relations. As in the case of the descent in symmetry from the L' levels, the subduced representations have to be multiplied by the pseudo-scalar representation to yield the final results in Eq. 36.

$$(1/2) \rightarrow E' \xrightarrow{\otimes A_2} E''$$
$$(3/2) \rightarrow U' \rightarrow U'$$
$$(5/2) \rightarrow E'' + U' \rightarrow E' + U' \qquad (36)$$

Correlation lines are drawn between strong and weak field schemes in accordance with the non-crossing rule. The E' and E'' levels, which belong to unique J' values, will not be affected by s.o.c. and therefore give rise to straight lines in the correlation diagram. The evolution of the U' levels is more intricate. As indicated before the U' levels of $^4S'$ and $^2D'$ parentage will interact with the $U'(^2P')$ component, but furthermore the composition of the two $U'(^2D')$ levels will vary as a function of the coupling conditions. In the weak s.o.c. limit these two levels separate as $U'(^2E_g) + U'(^2T_{1g})$ under the influence of configuration interaction with higher excited ligand field states. In the strong s.o.c. limit the U' levels separate as $U'(3/2) + U'(5/2)$. The two separations do not coincide but are related by a unitary transformation. The branching ratio's are as follows:

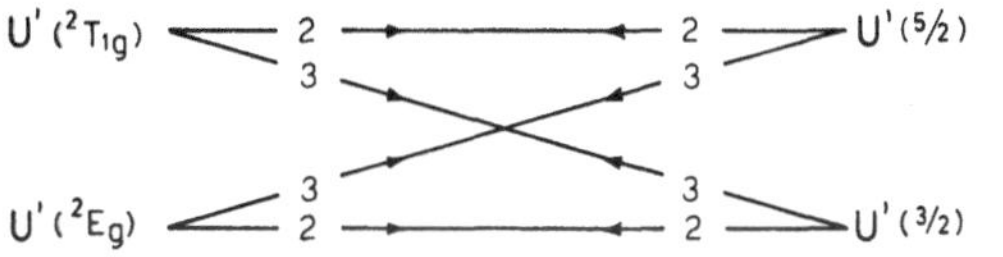

(37)

The ratio of 2:3 reflects the ratio of the orbital degeneracies of the two crystal field components.[1] According to this coupling scheme the crystal field parentage of the emitting U' level changes from pure ^2E_g in the weak s.o.c. case to a 3:2 predominance of $^2T_{1g}$ over ^2E_g in the strong coupling case. All Cr^{3+} complexes belong to the weak field coupling limits in view of the low s.o.c. constant for trivalent chromium ($\zeta = 273\ \mathrm{cm}^{-1}$). Hence in order to explore the correlation diagram of Fig. 2 one has to turn to 4*d* and 5*d* transition metals, preferably in high oxidation states. In Table 8 we have gathered experimental zero-phonon lines from detailed spectroscopic measurements on d^3 complexes of three heavier transition-metals [34–37]. The Table also lists the calculated levels based on standard ligand-field calculations using accepted parameter values for spectrochemical strength (10 *Dq*), interelectronic repulsion (*B*, *C*) and spin-orbit coupling (ζ). The agreement between theory and experiment is usually within $100\ \mathrm{cm}^{-1}$. Of course more refined calculations could be performed involving rather delicate features such as anisotropic spin-orbit coupling or the Trees correction [37]. Reliable estimates of such corrections are difficult though, unless one can pinpoint a specific observable effect, such as a term splitting, which is directly proportional to the parameter under study. In the present case where such effects are absent detailed considerations of these additional interactions prove to be of little value [34].

Table 8. Transition energies of the doublet states for $4d^3$ and $5d^3$ complexes (in cm^{-1}; calculated values in parentheses)

Excited state	$[MoCl_6]^{3-}$ [a]	$[ReF_6]^{2-}$ [b]	IrF_6 [c]
$a\Gamma_8(^2E_g, {}^2T_{1g})$	9291 (9206)	9070 (9192)	6260 (6440)
$b\Gamma_8(^2E_g, {}^2T_{1g})$	9603 (9563)	10631 (10401)	8330 (8132)
$\Gamma_6(^2T_{1g})$	9706 (9742)	11200 (11140)	8860 (8507)
$\Gamma_7(^2T_{2g})$	14552 (14801)	17394 (17431)	12330 (13196)
$\Gamma_8(^2T_{2g})$	14477 (14621)	18667 (18761)	15160 (15826)

[a] Mo^{3+} in Cs_2NaYCl_6 at 1.6 K [34]; 10 *Dq* = 18800; B = 495, C = 1820, ζ = 600.
[b] Re^{4+} in Cs_2GeF_6 at 20 K [35]; 10 *Dq* = 34830, B = 566, C = 1795, ζ = 2953.
[c] IrF_6 vapor phase [36]; 10 *Dq* = 40551 [37], B = 310, C = 1472 [5], ζ = 3331 [37].

[1] Interestingly an analogous result is obtained for the branching ratio of two W' spin-orbit levels resulting from a parent ^2F term in an icosahedral field [33].

Table 9. Contributions (in %) of the $(t_{2g})^3$ crystal field states to the eigenvectors of the lowest Γ_8 levels

Γ_8 level	$[MoCl_6]^{3-}$				$[ReF_6]^{2-}$				IrF_6			
	$^4A_{2g}$	^2E_g	$^2T_{1g}$	$^2T_{2g}$	$^4A_{2g}$	^2E_g	$^2T_{1g}$	$^2T_{2g}$	$^4A_{2g}$	^2E_g	$^2T_{1g}$	$^2T_{2g}$
$\Gamma_8(^4A_{2g})$	99.5	0.0	0.0	0.2	92.8	0.2	0.3	4.4	84.8	1.1	1.5	10.3
$a\Gamma_8(^2E_g, {}^2T_{1g})$	0.0	96.3	0.4	0.6	2.6	39.0	39.5	14.7	8.7	29.5	43.1	15.9
$b\Gamma_8(^2E_g, {}^2T_{1g})$	0.0	0.3	96.7	1.3	0.0	50.1	46.1	0.2	0.0	58.0	39.4	0.0
$\Gamma_8(^2T_{2g})$	0.3	0.8	1.4	89.2	3.1	6.8	11.4	73.8	5.0	8.9	13.9	70.3

In Table 9 we represent the composition of the U' eigenvectors coming out of the ligand field calculation. The three complexes are found to cover the entire range from weak to strong spin-orbit interactions. In the $[MoCl_6]^{3-}$ complex the lowest excited U' level is close to the pure $U'(^2E_g)$ starting level, indicating that this complex is an example of the weak coupling limit. Furthermore for this complex the sign of the $^2P'$ splitting is opposite to what is expected on the basis of intra-configurational interactions. Apparently this anomalous sign is caused by s.o.c. with the nearby $^4T_{2g}$ and $^4T_{1g}$ states belonging to the excited $(t_{2g})^2(e_g)^1$ configuration. The $[ReF_6]^{2-}$ case is intermediate in that the ^2E_g and $^2T_{1g}$ are both present almost to equal extent in the lowest excited U' level. Finally the IrF_6 complex, measured in the vapor phase, probably represents the most pronounced example of the strong coupling limit. In this case the ratio of ^2E_g over $^2T_{1g}$ in the emitting level is indeed close to 2:3. This level thus approaches the $U'(3/2)$ eigenvector. Likewise the next U' level is seen to adopt the composition of the $U'(5/2)$ eigenvector.

4.3 Anisotropic Spin-Orbit Coupling

In low-symmetry environments the octahedral U' levels undergo a splitting into two doubly degenerate sublevels. The mechanism of this splitting involves higher-order interactions since all first-order perturbation terms are forbidden by the internal selection rules of the half-filled shell. The splitting of the U' levels originating from the $^2P'$ and $^2D'$ terms yields information on the low symmetry components of the ligand field and will be discussed in the subsequent section. Here particular attention will be devoted to the zero-field splitting (zfs) of the $^4A_{2g}$ ground state in trigonal fields. In D_3 the $^4A_{2g}$ level is resolved into E' and E'' sublevels, comprising resp. the $M_S = \pm 1/2$ and $M_S = \pm 3/2$ components. The zfs is defined as $D(^4A_{2g}) = E(E') - E(E'')$. It can be measured by EPR or optical spectroscopy. In Table 10 some zfs values for octahedral complexes with D_3 (or C_3) microsymmetry are presented [38–43].

In their early study of the ruby spectrum Sugano and Tanabe [44] attributed the zfs to the trigonal anisotropy of the spin-orbit coupling. In a simplified form the anisotropic s.o.c. hamiltonian may be written as:

$$\mathscr{H}^{SOC} = \zeta_{\parallel} \mathscr{L}_z \mathscr{S}_z + \zeta_{\perp}(\mathscr{L}_x \mathscr{S}_x + \mathscr{L}_y \mathscr{S}_y) \tag{38}$$

Table 10. Zero field splitting (in cm^{-1}) of the ${}^4A_{2g}$ ground state in trigonal Cr(III) complexes

	$E(E') - E(E'')$	lattice	Ref.
$[Cr(en)_3]^{3+}$	0.00495	$[Cr(en)_3]Cl_3 \cdot NaCl \cdot 6H_2O$	[38]
$[CrO_6]$	0.38	ruby	[39,40]
$Cr(acac)_3$	1.20	$Al(acac)_3$	[41]
$[Cr(bipy)_3]^{3+}$	0.82	$[Cr(bipy)_3](PF_6)_3$	[42]
	0.8 à 0.9	amorphous hosts	[43]

This hamiltonian has cylindrical symmetry and may be used to introduce trigonal or tetragonal anisotropy, depending on whether the principal z axis is oriented along a C_3 or C_4 symmetry axis. The second-quantized form of the intra-t_{2g} part of this operator is given in Eq. 39.

$$\mathscr{L}_z\mathscr{S}_z = 1/2(-a^{\dagger}_{\alpha+1}a_{\alpha+1} + a^{\dagger}_{\alpha-1}a_{\alpha-1} + a^{\dagger}_{\beta+1}a_{\beta+1} - a^{\dagger}_{\beta-1}a_{\beta-1})$$

$$\mathscr{L}_x\mathscr{S}_x + \mathscr{L}_y\mathscr{S}_y = -\frac{1}{\sqrt{2}}(a^{\dagger}_{\alpha 0}a_{\beta+1} + a^{\dagger}_{\beta+1}a_{\alpha 0} + a^{\dagger}_{\beta 0}a_{\alpha-1} + a^{\dagger}_{\alpha-1}a_{\beta 0}) \tag{39}$$

Now this operator can directly be applied to the spherical eigenvectors of Table 2. For the $M_S = +3/2$ and $M_S = +1/2$ components of ${}^4A_{2g}$ one obtains:

$$\mathscr{H}^{SOC}|{}^4S' + 3/2\,0\rangle = -\zeta_{\perp}|{}^2P' + 1/2 + 1\rangle$$

$$\mathscr{H}^{SOC}|{}^4S' + 1/2\,0\rangle = -(\sqrt{2}\zeta_{\parallel}|{}^2P' + 1/2\,0\rangle + \zeta_{\perp}|{}^2P' - 1/2 + 1\rangle/\sqrt{3} \tag{40}$$

Hence the E' level ($M_S = \pm 1/2$) is connected to the ${}^2P'$ term via $\zeta_{\parallel}$ and $\zeta_{\perp}$ elements in a 2:1 ratio, while the E'' level ($M_S = \pm 3/2$) is connected via the $\zeta_{\perp}$ part of the hamiltonian. This anisotropy gives rise to the following second-order contribution to the zfs:

$$D({}^4A_{2g}) = 2/3(\zeta_{\perp}^2 - \zeta_{\parallel}^2)/(E({}^2P') - E({}^4S')) \tag{41}$$

A more general expression involving states of the $(t_{2g})^2(e_g)^1$ configuration has recently been presented by Dubicki [45]. It should be noted that Eq. 41 applies both to tetragonal and trigonal zfs. This is a consequence of the pseudo-spherical symmetry of the t_{2g}-shell and will be further elaborated in the subsequent section.

The result in Eq. 41 is reminiscent of the expression for anisotropic interelectronic repulsion presented in Sect. 3.2. However unlike in the latter case, no attempt was made to analyze the orbital mechanisms underlying the spin-orbit anisotropy. Instead it was shown later on that the zfs in ruby could be accounted for by a full ligand field calculation combining isotropic s.o.c. with a trigonal perturbation of the ligand field [46,47].

Macfarlane [40, 48] traced the resulting splitting back to third-order perturbation loops, quadratic in the isotropic s.o.c. constant and linear in the trigonal field parameter v'. This parameter describes the off-diagonal ligand field matrix element between t_{2g} and e_g orbitals. According to the Angular Overlap Model (AOM) of the ligand field v' is proportional to the geometric distortion of the octahedral frame [49]. Hence for a complex with a large trigonal distortion, such as the $[CrO_6]$ complex in ruby [47], a pronounced zfs is expected and indeed observed. In contrast in a complex which is nearly orthoaxial, such as $[Cr(en)_3]^{3+}$, the zfs ought to be very small, again in line with experiment (Cf. Table 10).

In the case of $Cr(acac)_3$ this analysis runs into serious difficulties. The six oxygen ligators in this complex occupy positions which are very near to the vertices of a regular octahedron [50]. Accordingly standard ligand field calculations are quite unable to rationalize the large splitting which is found experimentally [51]. Even the sign of the calculated zfs appears to be wrong. In view of this discrepancy Atanasov and Schönherr [51] proposed to return to the original anisotropy hypothesis of Tanabe and Sugano [14].

Apparently the ups and downs of this hypothesis call for a comment. The anisotropic hamiltonian in Eq. 38 really must be looked upon as an effective operator which incorporates the mixing between the t_{2g} orbitals and other orbitals of a different nature. In the case of a trigonally distorted complex these other orbitals are *inter alia* the e_g orbitals on the metal. Hence rather than being opposed to it, Macfarlane's analysis [48] actually complies with the anisotropy hypothesis by offering an explicit mechanism underlying the apparent anisotropy. Similarly it may be assumed that in a complex such as $Cr(acac)_3$ which has low-lying LMCT states orbital mixing will occur between the t_{2g} orbitals and π-type orbitals on the ligand. It is likely that in this case the resulting anisotropy of the s.o.c. hamiltonian can be described explicitly by invoking higher-order perturbation loops involving the charge-transfer states. It may further be conjectured that a similar mechanism will contribute to the zfs in the related $[Cr(bipy)_3]^{3+}$ complex, which has prominent low-lying MLCT states. In principle ab initio calculations could help to clarify this issue by providing information on the degree of orbital mixing in these charge-transfer type complexes.

5 Static Ligand Field

In this section, we will be concerned with the low-symmetry components of the ligand field (LF) in hexacoordinated complexes. Cases of tetragonal, trigonal and orthorhombic symmetry breaking will be developed, but first attention will be devoted to the general selection rules governing these interactions.

5.1 Quasi-Spin and Pseudo-Cylindrical Selection Rules

Low-symmetry LF operators are time-even one-electron operators that are non-totally symmetric in orbit space. They thus have quasi-spin $K = 1$, implying that the only allowed matrix elements are between $^2P'$ and $^2D'$ (Cf. Eq. 28). Interestingly in complexes with a trigonal or tetragonal symmetry axis a further selection rule based on the angular momentum theory of the shell is retained. Indeed in such complexes two t_{2g}-orbitals will remain degenerate. This indicates that the intra-t_{2g} part of the LF hamiltonian has pseudo-cylindrical $D_{\infty h}$ symmetry. As a result the $^{2S+1}L'$ terms are resolved into pseudo-cylindrical $^{2S+1}\Lambda'$ levels ($\Lambda' = 0, 1, \ldots, L'$). It is convenient to orient the z axis of quantization along the principal axis of revolution. In this way each Λ' level comprises the $M_{L'} = \pm\Lambda'$ components of the L' manifold. In a pseudo-cylindrical field only levels with equal Λ' are allowed to interact, in accordance with the pseudo-cylindrical selection rule:

$$\Delta\Lambda' = 0 \tag{42}$$

In reality the cylindrical LF has to be added to the octahedral interaction which lifts the degeneracy of the $^2D'$ term as described in Sect. 3.1. When combining these two perturbations it matters whether the cylindrical field coincides with the trigonal or with the tetragonal directions of the octahedron, because the remaining D_{3d} or D_{4h} subgroups are not equivalent. Two different spectral patterns thus will result from this superposition of octahedral and cylindrical ligand fields.

5.2 Tetragonal Fields

Most examples of tetragonal complexes are mono- and disubstituted chromium(III) amine complexes of the type CrN_5X, *trans*-CrN_4XY, *cis*-CrN_4X_2, where X and Y are acido ligands such as F^-, Cl^-, NCS^-, OH^-, H_2O and the N are nitrogen donors such as NH_3, ethylenediamine or pyridine. Although the actual point group symmetry of these complexes may be quite low the effective holohedron symmetry is always D_{4h}, due to the presence of two identical coordination axes. According to the AOM the orbital energies of the t_{2g} orbitals depend on the average π parameters, π_{eq} and π_{ax}, of the ligands on the equatorial (x, y) and axial (z) coordination sites [52]. In second-quantized form the traceless tetragonal LF potential is given by:

$$\mathscr{V}^{D_4} = 2/3(\pi_{eq} - \pi_{ax})(2a^\dagger_{\alpha 0}a_{\alpha 0} + 2a^\dagger_{\beta 0}a_{\beta 0} - a^\dagger_{\alpha+1}a_{\alpha+1} - a^\dagger_{\alpha-1}a_{\alpha-1} - a^\dagger_{\beta+1}a_{\beta+1} - a^\dagger_{\beta-1}a_{\beta-1}) \tag{43}$$

This hamiltonian has cylindrical symmetry around the axis of quantization. It lifts the degeneracy of the spherical $^2P'$ and $^2D'$ terms to yield the following components:

$$^2P' \rightarrow {}^2\Sigma_g'^- + {}^2\Pi_g'$$
$$^2D' \rightarrow {}^2\Sigma_g'^+ + {}^2\Pi_g' + {}^2\Delta_g' \qquad (44)$$

Note that we have chosen to use primed symbols to emphasize the shell-theoretical nature of these labels, as opposed to true spatial symmetry assignments. The corresponding tetragonal symmetries can easily be found from the standard $D_{\infty h} \downarrow D_{4h}$ subduction relations, keeping in mind that the results must be multiplied by the pseudoscalar B_{1g} representation of D_{4h}. One thus obtains:

$$\Sigma_g'^+ \rightarrow A_{1g} \xrightarrow{\otimes B_{1g}} B_{1g}$$
$$\Sigma_g'^- \rightarrow A_{2g} \rightarrow B_{2g}$$
$$\Pi_g' \rightarrow E_g \rightarrow E_g$$
$$\Delta_g' \rightarrow B_{1g} + B_{2g} \rightarrow A_{1g} + A_{2g} \qquad (45)$$

These levels may be related to the doublet states in the unperturbed octahedron according to the following branching scheme:

$$^2E_g \rightarrow {}^2A_{1g} + {}^2B_{1g}$$
$$^2T_{1g} \rightarrow {}^2A_{2g} + {}^2E_g$$
$$^2T_{2g} \rightarrow {}^2B_{2g} + {}^2E_g \qquad (46)$$

In Fig. 3 we represent the energy evolution of these doublets as a function of increasing tetragonal perturbation. At the far right of the diagram the levels are correlated to the cylindrical splitting pattern. As can be seen from the diagram the only energetic effect of the perturbation is the second-order interaction between the two levels of Π' symmetry. This is in line with the selection rule of Eq. 42. Furthermore in the pseudo-cylindrical limit the $^2\Sigma_g'$ and $^2\Delta_g'$ levels are degenerate. This degeneracy is partially lifted due to the different octahedral parentage of both levels. A further resolution of the $^2A_{1g}$ and $^2B_{1g}$ components may be induced by anisotropic electron repulsion effects as discussed in Sect. 3.2.

The most striking aspect of the diagram is the crossing of the lowest doublet states. In complexes with low tetragonal fields ($|\pi_{eq} - \pi_{aq}| < 1000\ \mathrm{cm}^{-1}$) the lowest lying levels are the quasi-degenerate $^2A_{1g}$ and $^2B_{1g}$ components of ^2E_g parentage. At higher fields the nature of the lowest doublet switches over to the degenerate ^2E_g components of $^2T_{1g}$ parentage. Since emission originates from the lowest lying excited level of the manifold two classes of emitters are to be found, commonly denoted as ^2E and ^2T emitters. After examining over forty quadrate chromium(III) amine complexes Forster, Rund and Fucaloro [53] formulated an important emission rule: ^2E emitters produce sharp emission spectra in the 650–710 nm region, while ^2T emitters are characterized by broad spectra, usually shifted to longer wavelengths. It has also been found [54] that the two types of emitters have different nonradiative relaxation rates and

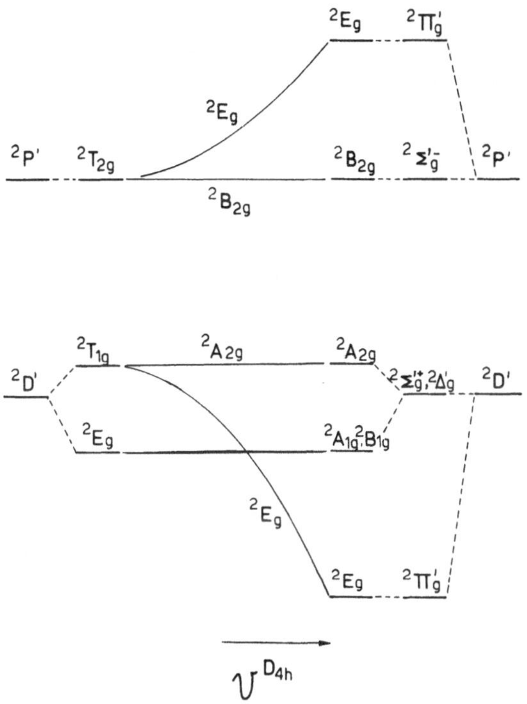

Fig. 3. Correlation diagram of the doublet states relating weak (*left*) and strong (*right*) tetragonal perturbations. In the strong field limit the splitting pattern is determined by pseudo-Λ' quantum numbers. Note the crossing point on the lowest energy curve

different solvent effects. ^2E emitters apparently are rather imperturbable entities in that their relaxation rates are confined to a narrow range, $(0.9\text{–}3.5) \times 10^4\,s^{-1}$ at 77 K in glassy solutions, while solvent changes have only minor effects. In contrast the decay rates for ^2T emitters may vary over a considerable range and the emission shifts to shorter wavelengths in hydroxylic solvents [55].

These differences between the two types of emitters have been related to configurational differences of the emitting states [52]. The $^2A_{1g} + {}^2B_{1g}$ levels are not affected by the tetragonal field and therefore retain the isotropic charge distribution of one electron per t_{2g} orbital which is characteristic of the parent half-filled shell states [56]. In contrast the emitting level of a ^2T emitter is composed of a mixture of the two $^2\Pi'_g$ functions of $^2P'$ and $^2D'$ origin. In consequence there is a polarization of the charge distribution, the lower lying t_{2g} orbitals being more populated at the expense of the higher lying ones. This transfer of charge affects the metal-ligand bond strengths, induces Jahn-Teller instability and influences interactions with the solvent. As a result the emission spectrum acquires to a certain extent the typical features of the broad-hand emission spectra which are characteristic of the photoactive quartet states. At one time ^2T emission was even mistakenly identified as delayed fluorescence [57].

In the perspective of shell-theoretic selection rules the two types of emission clearly expose a difference in quasi-spin characteristics. The ^2E-emitting levels are $^2D'$ components and thus pure quasi-spin singlets (Cf. Table 3). This makes them immune for all time-even one-electron operators in accordance with the stringent quasi-spin selection rules. In contrast in the ^2T emitters a strong tetragonal perturbation has been invested to form a mixture of a quasi-spin singlet ($^2D'$) and triplet ($^2P'$). To this mixture the quasi-spin selection rules no longer apply, and all sorts of perturbing interactions may develop.

From the point of view of technological applications [58], complexes with a quadrate field in the border zone between ^2E and ^2T emitters are perhaps of particular interest. Indeed for such complexes very slight changes in the environment may easily lead to dramatic changes in the emission spectra. This effect has indeed been demonstrated by Fucaloro et al. [55] for the *cis*-dihydroxo complexes *cis*-$[Cr(NH_3)_4(OH)_2]^+$ and *cis*-$[Cr(en)_2(OH)_2]^+$.

5.3 Trigonal Fields

As we have indicated before tetragonal and trigonal fields are both examples of cylindrical fields and thus may be treated on equal footing, provided the axis of quantization is oriented along the corresponding axis of revolution in the octahedral frame. In the preceding orbital expressions (Eq. 9) the standard tetragonal xyz frame was adopted with the z axis along a fourfold axis of the octahedron. In Fig. 4 a convenient trigonal coordinate frame, labeled $x'y'z'$, is presented with the z' axis along the C_3^{111} and the x' axis along a C_2 axis in between the x and y directions of the original frame. The C_3 and C_2 operators generate a D_3 group, which is the common point group symmetry of trischelated complexes. The t_1 components t_1x', t_1y', t_1z' may immediately be obtained by vector rotation of the t_1 components in the original frame:

$$|t_1x'\rangle = \frac{1}{\sqrt{2}}(|t_1x\rangle - |t_1y\rangle)$$

$$|t_1y'\rangle = \frac{1}{\sqrt{6}}(-2|t_1z\rangle + |t_1x\rangle + |t_1y\rangle)$$

$$|t_1z'\rangle = \frac{1}{\sqrt{3}}(|t_1x\rangle + |t_1y\rangle + |t_1z\rangle) \tag{47}$$

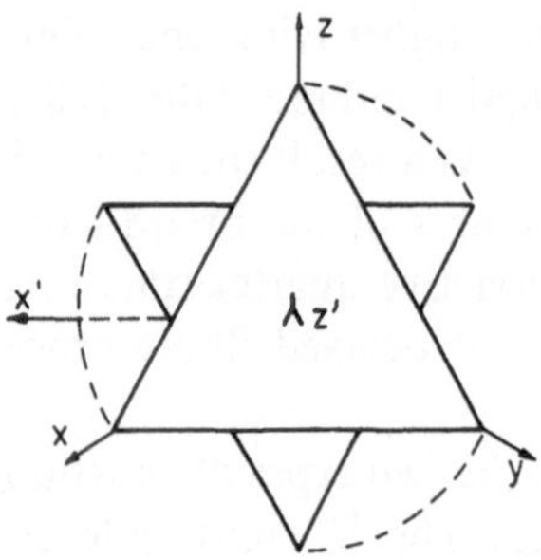

Fig. 4. Relationship between tetragonal (xyz) and trigonal ($x'y'z'$) coordinate frames in a trischelated complex

Starting from this rotated set complex orbitals and $(t)^3$ multiplet operators may be constructed in a way which is entirely analogous to the treatment of Sect. 2. Hence the multiplets in Table 2 can be used equally well for trigonal complexes, keeping in mind that the axis of quantization is now the z' axis. This implies that the subduction rules for real components in Eq. 15 have to be replaced by the appropriate $SO_3 \downarrow O \downarrow D_3$ subduction rules. In order to obtain the real forms of the $(t_2)^3$ basis functions the resulting expressions have to be multiplied once again by the pseudoscalar quantity of A_2 symmetry. The appropriate product rules have been given by Ballhausen [59]. For the individual orbital functions one obtains:

$$|t_2 a_1\rangle = A_2 |t_1 z'\rangle$$
$$|t_2 e_\theta\rangle = -A_2 |t_1 y'\rangle$$
$$|t_2 e_\varepsilon\rangle = A_2 |t_1 x'\rangle \tag{48}$$

Here a_1 and e are representations of D_3. The θ and ε components are resp. symmetric and antisymmetric under the C_2 operator along the x' direction. Combining Eqs. 47 and 48 with Eq. 9 then yields the conversion formula for the real t_{2g} orbitals in trigonal and tetragonal frames:

$$|t_2 a_1\rangle = \frac{1}{\sqrt{3}}(|t_2 yz\rangle + |t_2 xz\rangle + |t_2 xy\rangle)$$
$$|t_2 e_\theta\rangle = \frac{1}{\sqrt{6}}(2|t_2 xy\rangle - |t_2 yz\rangle - |t_2 xz\rangle)$$
$$|t_2 e_\varepsilon\rangle = \frac{1}{\sqrt{2}}(|t_2 yz\rangle - |t_2 xz\rangle) \tag{49}$$

The corresponding d-functions may also be expressed in the trigonal frame as follows:

$$|t_2 a_1\rangle = d_{z'^2}$$
$$|t_2 e_\theta\rangle = \frac{1}{\sqrt{3}}(d_{y'z'} - \sqrt{2} d_{x'^2-y'^2})$$
$$|t_2 e_\varepsilon\rangle = \frac{1}{\sqrt{3}}(\sqrt{2} d_{x'y'} - d_{x'z'}) \tag{50}$$

Pictures of the trigonal orbitals may be found in the literature [52, 60].

In classical crystal field theory [2] the trigonal field is parametrized by means of two independent parameters v' and v describing resp. the interaction between the t_{2g} and e_g shells and the splitting of the t_{2g} shell. The v' parameter was already discussed in Sect. 4.3 in connection with the trigonal zfs of the $^4A_{2g}$ ground state. Here special attention will be devoted to the v parameter which seems to dominate the doublet splittings, especially in orthoaxial trischelated complexes such as $Cr(acac)_3$ where v' is weak. The second-quantized form of the trigonal field due to the v operator may be written as follows:

$$\mathscr{V}^{D_3} = \frac{v}{3}(2a^{\dagger}_{\alpha 0}a_{\alpha 0} + 2a^{\dagger}_{\beta 0}a_{\beta 0} - a^{\dagger}_{\alpha+1}a_{\alpha+1}$$
$$- a^{\dagger}_{\alpha-1}a_{\alpha-1} - a^{\dagger}_{\beta+1}a_{\beta+1} - a^{\dagger}_{\beta-1}a_{\beta-1}) \quad (51)$$

where the orbital and spin labels refer to the z' axis of quantization. This hamiltonian is entirely analogous to the tetragonal hamiltonian in Eq. 43 and thus has pseudo-cylindrical symmetry around the z' axis. Hence the $^2P'$ and $^2D'$ manifolds will be resolved into cylindrical $^2\Lambda'$ levels exactly as in the previous case. Furthermore ligand field interactions will be limited to the off-diagonal elements connecting the $^2\Pi'$ levels in line with the cylindrical selection rules. The corresponding trigonal symmetries can be found from the standard $D_{\infty h} \downarrow D_3$ subduction relations, keeping in mind that the results must be multiplied by the pseudoscalar A_2 representation of D_3.

$$\Sigma_g^{\prime +} \rightarrow A_1 \xrightarrow{\otimes A_2} A_2$$
$$\Sigma_g^{\prime -} \rightarrow A_2 \rightarrow A_1$$
$$\Pi_g' \rightarrow E \rightarrow E$$
$$\Delta_g' \rightarrow E \rightarrow E \quad (52)$$

On the other hand the octahedral levels contain the following trigonal symmetry species:

$$^2E_g \rightarrow {}^2E$$
$$^2T_{1g} \rightarrow {}^2A_2 + {}^2E$$
$$^2T_{2g} \rightarrow {}^2A_1 + {}^2E \quad (53)$$

In Fig. 5 we represent the doublet energies as a function of increasing trigonal perturbation. At the far right the levels are correlated to the cylindrical splitting pattern. The main difference between the trigonal and tetragonal correlation diagrams is the absence of a curve crossing in the trigonal case due to the avoided crossing rule for equisymmetric states [52]. Hence as far as the ^2E components of the $^2D'$ term are concerned there is a gradual change of their composition as a function of increasing trigonal perturbation, corresponding to a change from octahedral coupling conditions on the left to cylindrical coupling conditions on the right. The corresponding branching ratios are as follows:

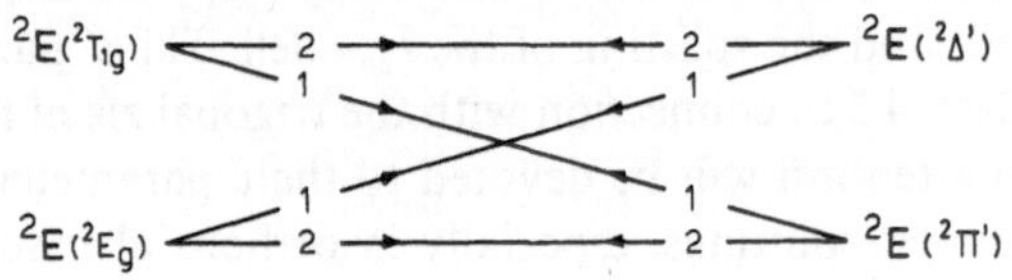

(54)

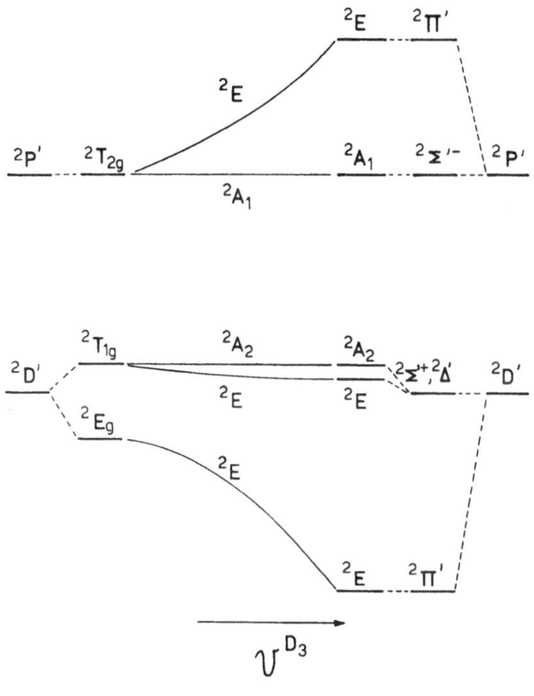

Fig. 5. Correlation diagram of the doublet states relating weak (*left*) and strong (*right*) trigonal perturbations. In the strong field limit the splitting pattern is determined by pseudo-Λ' quantum numbers. Note the absence of a crossing point in the lowest energy curve

According to this scheme the parentage of the lower lying ^2E level changes from pure ^2E_g in weak trigonal fields to a 2:1 ratio of ^2E_g versus $^2T_{1g}$ in strong fields. There is a concomitant rise of the $^2P'$ character of the emitting level, due to the interaction between $^2\Pi'(^2D')$ and $^2\Pi'(^2P')$.

As a result of the avoided crossing, all trigonal emitters essentially belong to one and the same emission class [52]. In this class there will be a gradual change of emission characteristics as a function of the trigonal field. It is predicted that stronger fields will weaken the quasi-spin selection rule which protects the emitting levels from surrounding nuclear perturbations. However according to Forster there is not yet a clear experimental verification of this prediction [6].

Spin-orbit coupling in conjunction with the trigonal field leads to a zero-field splitting of the ^2E levels. In the strong field limit the spin-orbit levels can be obtained by vector addition of cylindrical orbital and spin momenta. Hence the $^2\Pi'$ state will give rise to $^2\Pi'_{3/2}$ and $^2\Pi'_{1/2}$ components, comprising resp. the $|^2D' \pm 1/2 \pm 1\rangle$ and $|^2D' \pm 1/2 \mp 1\rangle$ functions. The trigonal symmetries of these functions are as follows:

$$\begin{aligned} &{}^2\Pi'_{1/2} \rightarrow E' \xrightarrow{\otimes A_2} E' \\ &{}^2\Pi'_{3/2} \rightarrow E'' \rightarrow E'' \end{aligned} \tag{55}$$

Both $\mathscr{V}^{D_3}$ and $\mathscr{H}^{SOC}$ will connect these components to the $^2P'$ term. Using the second-quantized form of these operators, as specified in Eqs. 38 and 51, one obtains for the $^2\Pi'_{3/2}$ levels:

$$(\mathscr{V}^{D_3} + \mathscr{H}^{SOC})|^2D' + 1/2 + 1\rangle = (v - 1/2\zeta_{\|})|^2P' + 1/2 + 1\rangle, \quad (56a)$$

and likewise for the $^2\Pi'_{1/2}$ level:

$$(\mathscr{V}^{D_3} + \mathscr{H}^{SOC})|^2D' - 1/2 + 1\rangle = (v + 1/2\zeta_{\|})|^2P' - 1/2 + 1\rangle + 1/\sqrt{2}\zeta_{\perp}|^2P + 1/20\rangle \quad (56b)$$

In second-order perturbation theory the cross terms resulting from Eq. 56 contribute to the zfs of the $^2\Pi'(^2D')$ level:

$$E(E') - E(E'') = -2v\zeta_{\|}/(E(^2P') - E(^2D')) \quad (57)$$

In small trigonal fields this result must be attenuated by a factor 2/3 to account for the branching ratio of the coupling scheme in Eq. 54. In this way the zfs becomes $-4/3v\zeta/(E(^2P') - E(^2D'))$, which is the well known result first obtained by Sugano and Tanabe [44]. Most importantly the derivation shows that the zfs of the emitting level is a linear function of v. It can thus be used to determine the sign of the trigonal splitting of the t_{2g} shell [61].

Examples of trigonal Cr(III) complexes were listed in Table 10. Multiplet splittings of these complexes have been commented upon by Dubicki [45]. In recent years special attention has been devoted to trischelated complexes containing unsaturated ligands such as oxalate, acetylacetonate or bipyridine. It was conjectured by Orgel [62] as early as 1961 that the trigonal splitting of the t_{2g} shell in these complexes could arise from specific electronic interactions between the frontier π-orbitals on the conjugated bidentate and the metal t_{2g}-orbitals. This idea is unusual in the framework of classical ligand field theory since it assumes interference between the fields of bridged ligator atoms. The term phase-coupling was introduced to describe this specific through-bridge electronic interaction [63]. A model was developed which incorporates the effect of phase-coupling in the AOM formalism [64, 65]. Two parameters, e_{ψ} and e_{χ}, corresponding resp. to in-phase and out-of-phase coupling, were defined as illustrated in Fig. 6. For a trischelated complex application of this model yields the following expression for the trigonal splitting of the t_{2g}-shell:

$$v = 3/2(e_{\chi} - e_{\psi}) \quad (58)$$

In the case of $Cr(acac)_3$ a large positive zfs of about 250 cm^{-1} is observed for the lowest ^2E level [66]. This points to a negative v parameter of about $-2100\ cm^{-1}$ [51]. A similar value is obtained from a parametric fit of the spin-allowed absorption bands in the visible [50]. According to molecular orbital considerations $acac^-$ has a prominent highest occupied orbital which provides an in-phase coupling of the p_{π} orbitals on the ligator atoms. If we neglect the e_{χ} contribution Eq. 58 yields an e_{ψ} value of $+1400\ cm^{-1}$, which indeed corresponds to the presence of strong in-phase-coupled oxygen donors. On the other hand in $[Cr(ox)_3]^{3-}$ the zfs of the ^2E state is much smaller ($20 \sim 40\ cm^{-1}$) and very much dependent on the surrounding lattice [67].

ψ-type

	d_{xz}	d_{yz}
d_{xz}	$1/2 e_\psi$	$1/2 e_\psi$
d_{yz}	$1/2 e_\psi$	$1/2 e_\psi$

χ-type

	d_{xz}	d_{yz}
d_{xz}	$1/2 e_\chi$	$-1/2 e_\chi$
d_{yz}	$-1/2 e_\chi$	$1/2 e_\chi$

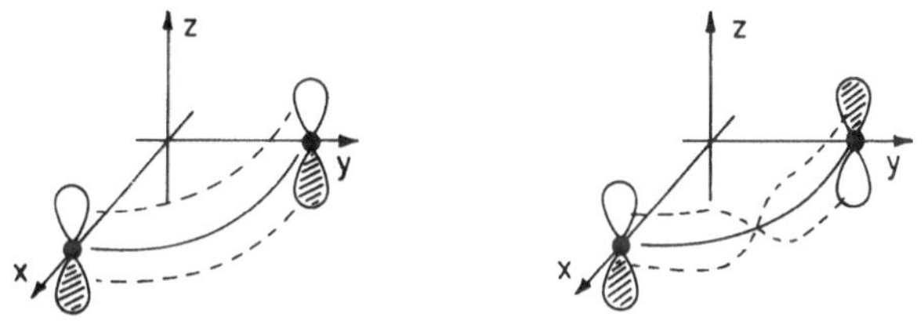

Fig. 6. In-phase (ψ) and out-of-phase (χ) coupling between the p_z orbitals on the ligator atoms of an unsaturated bidentate. The corresponding parameters e_ψ and e_χ are defined by the perturbational matrices of the d_{xz} and d_{yz} orbitals for the standard ligator positions as indicated

A study of the spin-allowed bands in this complex has revealed that there is no observable phase-coupling effect [50]. According to Atanasov et al. [50] this must be explained by the near-degeneracy of donor levels of ψ and χ symmetry type. We consider it also conceivable that coordination of the metal ion polarizes the charge distribution in the ligand chain as depicted below:

I II

In bond isomer II conjugation is displaced towards the outer oxygens, thus removing phase-coupling from the ligator atoms.

Finally for the case of $[Cr(bipy)_3]^{3+}$ the situation remains unclear. Hückel theory [63] predicts a positive v parameter mainly as a result of an out-of-phase coupled π-acceptor interaction ($e_\chi < 0$). For the lowest ^2E level a zfs of $-19.5\ cm^{-1}$ has been reported [42]. While the sign of this splitting is in agreement with the Hückel results, the magnitude is deceivingly small. It might be that in this case there is a sizeable geometric distortion which counteracts the phase-coupling effect.

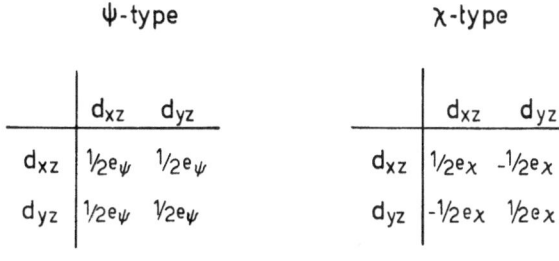

5.4 Orthorhombic Fields

In orthorhombic fields, no degeneracies of the t_{2g}-shell survive. As a result the pseudo-spherical properties of the shell can no longer be used. Only the quasi-spin selection rule, limiting all orthorhombic interactions to the $^2P' - {}^2D'$ off-diagonal elements, remains. To see which elements will be allowed the orthorhombic symmetries of the doublet states have to be determined. Interestingly two distinct embeddings of D_{2h} in O_h are possible. They have been labeled D_{2h}^{I} and D_{2h}^{II} and are presented in Fig. 7. In case I the C_2 axes are along the coordinate axes. This case corresponds to the holohedron symmetry of complexes with three different coordinate axes such as *cis*-CrN_4XY or *mer*-CrA_3B_3. The $O_h \downarrow D_{2h}^{\mathrm{I}}$ subduction yields the following branching:

$$
\begin{aligned}
&{}^2D': \begin{cases} {}^2E_g \to 2\,{}^2A_g \\ {}^2T_{1g} \to {}^2B_{1g} + {}^2B_{2g} + {}^2B_{3g} \end{cases} \\
&{}^2P': {}^2T_{2g} \to {}^2B_{1g} + {}^2B_{2g} + {}^2B_{3g}
\end{aligned}
\tag{59}
$$

According to this scheme the orthorhombic symmetries of $^2T_{1g}$ and $^2T_{2g}$ coincide. Application of the quasi-spin selection principle then leads to the conclusion that the orthorhombic perturbation will induce a mixing of the $^2T_{1g}$ and $^2T_{2g}$ components, leaving the ^2E_g components unaffected. Forster's emission rule [53] thus suggests that most of these complexes may be characterized as ^2E emitters, ^2T emission only being observable in strongly perturbed complexes. Detailed spectroscopic studies of D_{2h}^{I} complexes are absent from the literature. Recently though Choi and Hoggard [68] studied the sharp-line spectrum of a mer-type Cr(III) complex of glycylglycinate which has approximate D_{2h}^{I} symmetry. Calculations showed that the lowest doublet components were of predominant ^2E_g parentage. The splitting of the two ^2E_g components amounts to 200 cm^{-1}. The splitting mechanism is probably similar to Schmidtke's differential π-expansion model [30] for the tetragonal ^2E_g splitting (Cf. Sect. 3.2).

Case II has two C_2 axes in between the coordinate axes. It is exemplified by bischelated complexes of the type *trans*-$Cr(L\text{–}L)_2X_2$ where L–L symbolizes

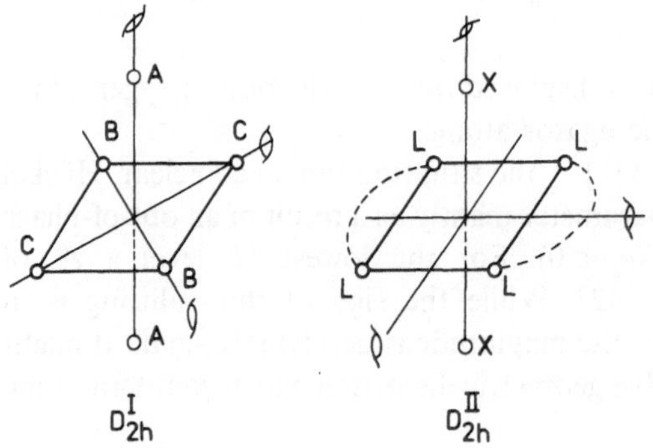

Fig. 7. Two types of $D_{2h} \subset O_h$ embedding

a bidentate ligand. The $O_h \downarrow D_{2h}^{\mathrm{II}}$ symmetry correlations are as follows:

$$^2D': \begin{cases} ^2E_g \to {}^2A_g + {}^2B_{1g} \\ ^2T_{1g} \to {}^2B_{1g} + {}^2B_{2g} + {}^2B_{3g} \end{cases}$$

$$^2P': {}^2T_{2g} \to {}^2A_g + {}^2B_{2g} + {}^2B_{3g} \tag{60}$$

In this case the ^2E_g and $^2T_{2g}$ states have a ^2A_g component in common. As a result there will be a direct quasi-spin and symmetry allowed interaction between ^2E_g and $^2T_{2g}$. This leads to an emitting level of $^2A_g(^2E_g)$ character which is perturbed by the orthorhombic field. The relevant interaction is proportional to the LF matrix element between d_{xz} and d_{yz} orbitals (coordinate frame as in Fig. 7). In the additive AOM this element is negligible unless the bite angle strongly deviates from 90°. Interestingly phase-coupling between the ligator atoms of the chelating ligand will enhance precisely the desired matrix element [63]. Indeed according to the model of phase-coupling (see Sect. 5.3) one has for a *trans*-$Cr(L-L)_2X_2$ complex with unsaturated L–L bidentates [64]:

$$\langle d_{xz} | \mathscr{V}^{D_{2h}} | d_{yz} \rangle = e_\psi - e_\chi \tag{61}$$

One thus would predict that such complexes are characterized by a pronounced orthorhombic perturbation of the emitting state, which thereby loses its parent quasi-spin structure. *Possibly this is the most efficient mechanism to make an emitting doublet state with a hole in the t_{2g} shell.* Interesting photophysical and even photochemical properties might result. In any case such complexes should constitute sensitive probes for the phase-coupling ligand properties.

So far detailed spectroscopic investigations of D_{2h}^{II} complexes have been limited to the *trans*-$[Cr(ox)_2(py)_2]^-$ complex, studied by Schönherr and Degen [69]. Apparently in this case there is no evidence for a pronounced orthorhombic field. This should not be surprising in view of the faint phase-coupling strength of the oxalate bridge. The question as to whether the phase-coupling can exert a pronounced influence on the emitting properties of Cr(III) complexes remains largely open.

6 Zeeman Interactions

The Zeeman hamiltonian is a one-electron time-odd operator. Of all operators considered it is the only one transforming as a quasi-spin singlet. It thus obeys the $\Delta Q = 0$ selection rule, unlike the ligand field and spin-orbit interactions which follow the $\Delta Q = 1$ selection rule. The consequence of this is a dramatic change of interaction pattern. Till now all interactions were concentrated in the off-diagonal elements causing a mutual repulsion between states of different quasi-spin character. In contrast the Zeeman perturbation resides inside the multiplets, giving rise to a linear state specific splitting. A detailed study of the corresponding g values may provide information on the role of higher-order ligand field and spin-orbit contributions. In this section we will especially be concerned with the general symmetry aspects of the Zeeman interactions in the spin-orbit levels of the $(t_{2g})^3$ states. The three spin representations E', E'', and U' of cubic symmetry will be discussed. The key concept of the investigation is the isotropy of the effect in a magnetic field of arbitrary orientation.

6.1 The Cubic E' and E'' Representations

The cubic E' and E'' representations (Γ_6 and Γ_7, respectively, in the Bethe notation) occur in the spin-orbit decomposition of the ${}^2T_{1g}$ and ${}^2T_{2g}$ terms (see Fig. 2). A general analysis of the Zeeman effect in such states is due to Yeakel [70]. The MCD and MCPL techniques in principle offer a method to determine the magnitude and sign of the corresponding g values. Such studies have indeed been performed on Cr^{3+} and Mo^{3+} ions doped in an elpasolite lattice [17, 34, 71]. In this centrosymmetric environment the electronic origins are very weak and most intensity is concentrated in vibronically induced electric dipole transitions. The magneto-optical measurements have mainly been used to assign the contributing modes. A detailed quantitative examination of the excited state g values and their dependence on the orientation of the magnetic field has not yet been undertaken.

Group theory predicts that time-odd interactions for double-valued representations must transform as the symmetrized direct square. For E' and E'' representations one has:

$$[E' \times E'] = T_1$$
$$[E'' \times E''] = T_1 \tag{62}$$

T_1 indeed corresponds to the symmetry of a magnetic field in an octahedron. As a result both E' and E'' levels will exhibit a linear Zeeman effect. On a more fundamental level the two components of the E' state, say $E'\alpha$ and $E'\beta$, may be looked upon as the spinor components of a vector space. The corresponding SU(2) symmetry group contains the octahedral double group as a subgroup. The fundamental representation of SU(2) indeed subduces E' in O^*. Furthermore the symmetrized direct square of the spinor yields the fundamental vector of the space, transforming as T_1 in O. In a second-quantized form the two components of E' may be generated by $b^\dagger_{E'\alpha}$ and $b^\dagger_{E'\beta}$ creation operators (not to be confused with the $a^\dagger$ creation operators of the constituent t_{2g} spin orbitals). A tensor form such as $(b^\dagger b)^1$ then indicates a coupling of the spinor creation and annihilation to a resultant vector of angular momentum 1. The Zeeman hamiltonian, considered by Yeakel [70], may be written as a scalar product of this coupled quantity with the magnetic field vector H.

$$\mathscr{H}^{ZE} = J^{E'}(b^\dagger b)^1 \cdot H \tag{63}$$

Here $J^{E'}$ symbolizes a coupling constant, corresponding to the g constant of the spin hamiltonian formalism. Being a scalar product of two vectors the hamiltonian will be invariant under any simultaneous rotation of both. As a result the linear Zeeman effect of E' states is predicted to have SO(3) symmetry, although the environment only has cubic symmetry.

An analogous treatment can be applied to the E'' representations, since E' and E'' are associated representations:

$$E'' = A_2 \times E' \tag{64}$$

The situation is extremely similar to the construction of a pseudo-angular momentum theory for t_2 states based on the associated t_1 representation (cf. Eq. 8). In the present case the Zeeman hamiltonian for E'' will be identical to the hamiltonian for E' since it contains the square of the pseudoscalar A_2 which is equal to $+1$. As a result both E' and E'' states will exhibit an isotropic linear Zeeman effect. In tetragonal or trigonal symmetries the SO(3) symmetry of $\mathscr{H}^{ZE}$ will of course be broken. Large deviations from isotropy are not to be expected though since the subduced representations remain doubly degenerate. This is a distinct feature of states with cubic $E'(^2T_{1g})$ or $E''(^2T_{2g})$ parentage, to which we will return in the next section.

6.2 The Cubic U' Representation

The Zeeman effect in the cubic U' representation (Γ_8 in the Bethe notation) is quite exceptional in that the symmetrized square of U' contains the T_1 symmetry of the magnetic field twice.

$$[U' \times U'] = A_2 + 2T_1 + T_2 \tag{65}$$

The general Zeeman hamiltonian thus will contain two independent g values. A convenient separation of the product multiplicity may be based on spherical coupling coefficients. To this aim the $U' \times U'$ product space is put into correspondence with the space that results from the addition of two $j = 3/2$ angular momenta. Their symmetrized square yields P and F products, which subduce T_1 and $A_2 + T_1 + T_2$ resp.

$$[3/2 \times 3/2] = P + F \tag{66}$$

The two T_1 products can thus be distinguished by a spherical coupling label as PT_1 and FT_1. Now let $c^\dagger$ create the four components of U'. The appropriately coupled quantities of T_1 symmetry read $(c^\dagger c)^{PT_1}$ and $(c^\dagger c)^{FT_1}$. The Zeeman hamiltonian for a U' representation can thus be written as follows:

$$\mathscr{H}^{ZE} = J^{PT_1}(c^\dagger c)^{PT_1} \cdot H + J^{FT_1}(c^\dagger c)^{FT_1} \cdot H \tag{67}$$

where the J's are two independent coupling constants. In principle this hamiltonian has a cubic anisotropy. However Yeakel [70] has found by explicit diagonalization that for certain ratios of the two J parameters the Zeeman splitting is isotropic with respect to the orientation of the magnetic field. This

Table 11. Isotropies of the Zeeman hamiltonian[a] for the cubic U' representation

	Normal quartet	Inside out quartet	Pseudo-doublet
J^{PT_1}/J^{FT_1}	1/0	3/4	− 1/2
s	1	1/9	− 1/3
$^{2S+1}\Gamma$	$^4A_1, {}^2T_1$	$^4A_2, {}^2T_2$	$^2E, {}^4E$

[a] The parameter s is the ratio of the two g-values as defined by Yeakel [70].

result has an interesting group theoretical background [72]. A shell transforming as U' may indeed be viewed [73] as the fundamental spinor of the five-dimensional rotation group SO(5). Of course the Zeeman hamiltonian cannot possibly exhibit this full symmetry since it does not involve a scalar product of vectors in SO(5) space. However a root diagram of SO(5) reveals that several intermediate SO(3) groups exist. Three of these can be represented by the Zeeman hamiltonian. They correspond to the three cases considered by Yeakel and are listed in Table 11. The first case with $J^{FT_1} = 0$ corresponds to the 'normal' Zeeman effect of a spin quartet. In this case the outer branches of the splitting pattern carry the $m_j = \pm 3/2$ components, corresponding to the $U'\kappa$ and $U'\lambda$ components in the Griffith notation [3]. This possibility will of course be realized for a $^4A_1$ state, but also for the U' level of any $^2T_1$ state. Indeed, in a $U'(^2T_1)$ level orbit and spin contributions are coupled in a parallel fashion giving rise to a genuine quartet splitting. However, if the orbit part is multiplied by the A_2 pseudoscalar, to yield $^4A_2$ or $^2T_2$, the U' components are interchanged in accordance with the product rules in Eq. 68.

$$
\begin{aligned}
|U'\mu\rangle &= A_2|U'\kappa\rangle \\
|U'\nu\rangle &= -A_2|U'\lambda\rangle \\
|U'\kappa\rangle &= -A_2|U'\mu\rangle \\
|U'\lambda\rangle &= A_2|U'\nu\rangle
\end{aligned}
\tag{68}
$$

As a result for such levels the inner and outer branches of the Zeeman profile will be interchanged with respect to the 'normal' quartet splitting. For this reason we have denoted this case as an 'inside out' or 'inverted' quartet [72]. The pseudo-scalar multiplication which relates the two types of quartet splitting corresponds to a Weyl reflection in the root diagram of SO(5). In the center of this reflection we meet a third case of Zeeman isotropy, which we have called a pseudo-doublet. Since this case is invariant under the Weyl reflection inner and outer branches of the splitting pattern will coalesce. The U' level thus will split in only two sublevels, each of which is pseudo-degenerate. Evidently this case is observed for any ^2E state, since the E-type orbital component of this state cannot contribute to the Zeeman splitting. It also occurs for a $U'(^4E)$ state.

The three possible types of Zeeman isotropies are thus encountered in the doublets of chromium(III) complexes. It must be emphasized that this is not a characteristic feature of a half-filled shell as such but only depends on the orbital symmetry of the doublet states under consideration (Cf. Table 11). In principle off-diagonal interactions between states of different orbital symmetry may remove the isotropy. Such effects have not yet been detected in the case of chromium(III).

A more drastic perturbation concerns the zero-field splitting of the U' levels in complexes with trigonal or tetragonal symmetry. This splitting completely quenches the Zeeman effect in a direction perpendicular to the main symmetry axis, thus yielding an apparent $g_\perp = 0$ result. This result is confirmed by several ligand field calculations at various levels of sophistication [40, 45, 51]. Hence if a trigonal or tetragonal level shows a substantial Zeeman splitting under a perpendicular magnetic field a cubic U' parentage must be excluded. Such level can only stem from the cubic $E'(^2T_{1g})$ or $E''(T_{2g})$ states (Cf. Sect. 6.1).

The Zeeman spectrum of ruby [39, 74] is in line with this expectation. For the $[Cr(bipy)_3]^{3+}$ complex measurements of $g_\perp$ have not yet been performed. On the other hand for the $Cr(acac)_3$ complex in alumina and gallium host lattices the emitting state was shown to have $|g_\perp| \approx 1.9$. This value was obtained using a scanning monochromator in combination with the EPR data of the ground state as a standard [41]. Later on direct excited state measurements were also carried out using ground state to doublet excited state pumping [75]. This result calls for a $E'(^2T_{1g})$ assignment of the emitting level in $Cr(acac)_3$, clearly in conflict with the accepted $U'(^2E_g)$ assignment (Cf. Sect. 5.3) of this level [51]. This is perhaps one of the most perplexing unsolved problems of Chromium(III) spectroscopy.

7 Conclusions

In this treatise we have reviewed the narrow line intercombination bands in the spectra of hexacoordinated Chromium(III) complexes from the perspective of an atomic-like shell approach. This implies that all interactions are represented by internal creation and annihilation operators of the shell states. Powerful shell symmetries such as angular momentum and quasi-spin then come into play. In a way the rich spectroscopy of the Chromium doublets highlights the various quantum characteristics associated with these symmetries. Obvious limitations of the shell approach are of course related to interactions which mix configurations and degrade the purity of the shell. This is especially important for the study of vibronic intensities of these spin and parity forbidden bands which must involve contributions from other configurations. As we have illustrated the stringent selection rules governing the internal interactions of the shell often provide an unexpected opportunity to observe in detail small subsidiary effects due to such other configurations.

Finally looking at the future, we express our hopes that inorganic chemists will succeed to insert d^n and f^n ions in the center of suitable icosahedral cages.

The large fourfold and fivefold degenerate shells of the icosahedral point group embody internal symmetries that are immensely richer than the ones we became acquainted with while studying the Chromium doublets.

Acknowledgements. Financial support from the Belgian Government (Programmatie van het Wetenschapsbeleid) and from the Belgian National Science Foundation (NFWO) is gratefully acknowledged. Thanks are due to L.G. Vanquickenborne of the Katholieke Universiteit Leuven for a discussion of the ab initio results of the Leuven Quantum Chemistry Group, and to G.E. Stedman and S.R. Mainwaring of the University of Canterbury, Christchurch, for communicating a preliminary report on half-filled shell states. Figure 1 was copied from the monograph by Schläfer and Gliemann [1] with permission from the publisher.

References

1. Schläfer HL, Gliemann G (1969) Basic principles of ligand field theory (Translated from the German by DF Ilten). Wiley-Interscience, London
2. Ballhausen CJ (1962) Introduction to ligand field theory. McGraw-Hill, New York
3. Griffith JS (1964) The theory of transition-metal ions, 2nd edn. Cambridge University Press, Cambridge
4. Sugano S, Tanabe Y, Kamimura H (1970) Multiplets of transition-metal ions in crystals. Academic, New York
5. Lever ABP (1984) Inorganic electronic spectroscopy, 2nd edn. Elsevier, Amsterdam
6. Forster LE (1990) Chem Revs 90: 331
7. See the contribution by P. Hoggard
8. Judd BR (1967) Second quantization and atomic spectroscopy. Johns Hopkins Press, Baltimore
9. Condon EU, Odabasi H (1980) Atomic structure. Cambridge University Press, Cambridge
10. McGlynn SP, Vanquickenborne L, Kinoshita M, Carroll DG (1972) Introduction to applied quantum chemistry. Holt, Rinehart and Winston, New York
11. Fano U, Racah G (1959) Irreducible tensorial sets. Academic Press, New York
12. Racah G (1943) Phys Rev 63: 367
13. Butler PH (1981) Point group symmetry applications. Plenum Press, New York
14. Wybourne BG (1991) In: Florek W, Lulek T, Mucha M (eds) Proceedings of the International School on Symmetry and Structural Properties of Condensed Matter, September 6–12 1990, Zajączkowo. World Scientific, Singapore, p 155
15. Ceulemans A (1984) Mededelingen van de Koninklijke Academie voor Wetenschappen, Letteren en Schone Kunsten van België, 46: 81
16. Stedman GE (1987) J Phys A 20: 2629
17. Schwartz RW (1976) Inorg Chem 15: 2817
18. Dubicki L, Ferguson J, van Oosterhout B (1980) J Phys C 13: 2791
19. Flint CD, Matthews AP (1974) J Chem Soc Faraday Trans 2, 70: 1301
20. Urushiyama A, Schönherr T, Schmidtke H-H (1986) Ber Bunsenges Phys Chem 90: 1188
21. Geiser U, Güdel HU (1981) Inorg Chem 20: 3013
22. Mukherjee RK, Bera Sc, Bose A (1972) J Chem Phys 56: 3720
23. Haspeslagh L (1984) PhD thesis. Katholieke Universiteit Leuven, Leuven
24. Vanquickenborne LG, Haspeslagh L, Hendrickx M, Verhulst J (1984) Inorg Chem 23: 1677
25. Vanquickenborne LG, Coussens B, Postelmans D, Ceulemans A, Pierloot K (1991) Inorg Chem 30: 2978
26. Pierloot K, Vanquickenborne LG (1990) J Chem Phys 93: 4154
27. Flint CD, Matthews AP (1973) J Chem Soc Faraday Trans 2, 69: 419
28. Lee K-W, Hoggard PE (1988) Inorg Chem 27: 907
29. Riesen H (1988) Inorg Chem 27: 4677
30. Schmidtke H-H, Adamsky H, Schönherr T (1988) Bull Chem Soc Japan 61: 59
31. Kibler M, Grenet G (1985) Int J Quant Chem 27: 213
32. Koide S, Pryce MHL (1959) Phil Mag 3: 607

33. Fowler PW, Ceulemans A (1993) Theor Chim Acta 86: 315
34. Stranger R, Moran G, Krausz E, Güdel H, Furer N (1990) Mol Phys 69: 11
35. Lomenzo J, Patterson H, Strobridge S, Engstrom H (1980) Mol Phys 40: 1401
36. Brand JCD, Goodman GL, Weinstock B (1971) J Mol Spectrosc 38: 464
37. Hoggard PE (1986) Coord Chem Revs 70: 85
38. McGarvey BR (1964) J Chem Phys 41: 3743
39. Sugano S, Tanabe Y (1958) Disc Farad Soc 26: 43
40. Macfarlane RM (1970) Phys Rev B 1: 989
41. Fields RA, Haindl E, Winscom CJ, Kahn ZH, Plato M, Möbius K (1984) J Chem Phys 80: 3082
42. Hauser A, Mäder M, Robinson WT, Murugesan R, Ferguson J (1987) Inorg Chem 26: 1331
43. Riesen H, Krausz E (1992) J Chem Phys 97: 7902
44. Sugano S, Tanabe Y (1958) J Phys Soc Japan 13: 880
45. Dubicki L (1991) Comments Inorg Chem 12: 35
46. Sugano S, Peter M (1961) Phys Rev 122: 381
47. Lee K-W, Hoggard PE (1990) Inorg Chem 29: 850
48. Macfarlane RM (1967) J Chem Phys 47: 2066
49. Schäffer CE (1967) Proc Roy Soc London A297: 96
50. Atanasov MA, Schönherr T, Schmidtke H-H (1987) Theor Chim Acta 71: 59
51. Atanasov M, Schönherr T (1990) Inorg Chem 29: 4545
52. Ceulemans A, Bongaerts N, Vanquickenborne LG (1987) Inorg Chem 26: 1566
53. Forster LS, Rund JV, Fucaloro AF (1984) J Phys Chem 88: 5012
54. Forster LS, Rund JV, Fucaloro AF (1984) J Phys Chem 88: 5017
55. Fucaloro AF, Forster LS, Glover SG, Kirk AD (1985) Inorg Chem 24: 4242
56. Ceulemans A, Beyens D, Vanquickenborne LG (1982) J Am Chem Soc 104: 2988
57. Kirk AD, Porter GB (1980) J Phys Chem 84: 887
58. Reisfeld R, Jørgensen CK (1988) Structure & Bonding 69: 63
59. Ballhausen CJ (1979) Molecular electronic structures of transition metal complexes. McGraw-Hill, New York
60. Dionne GF, Palm BJ (1986) J Magn Resonance 68: 355
61. Ceulemans A, Bongaerts N, Vanquickenborne LG (1987) In: Yersin H, Vogler A (eds) Proceedings of the seventh international symposium on the photochemistry and photophysics of coordination compounds, March 29-April 2 1987, Elmau. Springer, Berlin, p 31
62. Orgel LE (1961) J Chem Soc 3683
63. Ceulemans A, Dendooven M, Vanquickenborne LG (1985) Inorg Chem 24: 1153
64. Ceulemans A, Vanquickenborne LG (1990) Pure & Appl Chem 62: 1081
65. Schäffer CE, Yamatera H (1991) Inorg Chem 30: 2840
66. Schönherr T, Eyring G, Linder R (1983) Z Naturforsch 38a: 736
67. Zarić S, Niketić SR (1991) Polyhedron 10: 2673
68. Choi J-H, Hoggard PE (1992) Polyhedron 11: 2399
69. Schönherr T, Degen J (1990) Z Naturforsch 45a: 161
70. Yeakel WC (1977) Mol Phys 33: 1429
71. Denning RG (1989) In: Flint CD (ed) Vibronic processes in inorganic chemistry. Kluwer, Dordrecht, p 111 (Nato ASI Series C, vol 288)
72. Ceulemans A, Mys G, Wałcerz S (1993) New J Chem 17: 131
73. Judd BR (1977) Colloques Internationaux CNRS 255: 127
74. Trabjerg I, Güdel HU (1974) Acta Chemica Scandinavica A28: 8
75. Fields RA, Winscom CJ, Haindl E, Plato M, Möbius K (1986) Chem Phys Lett 124: 121

Vibrational Progressions in Electronic Spectra of Complex Compounds Indicating Strong Vibronic Coupling

H.-H. Schmidtke

Institut für Theoretische Chemie, Universität Düsseldorf, Universitätsstr.1, D-40225 Düsseldorf, FRG

Table of Contents

Broad bands in optical absorption and emission spectra originating from one or several closely lying electronic transitions are indications of strong vibronic coupling. In the case when the vibrational structure is resolved into a progression of individual bands, a vibronic analysis can be carried out which compares a theoretical line shape function with the intensity profile measured in the

Topics in Current Chemistry, Vol. 171

experiment. From this Franck-Condon type procedure the normal modes to which the transition is coupled are determined, furthermore the strength of coupling, the stabilization energy involved, the distortion of complex molecule in the excited state and other information can be obtained. The theory derived on a semiclassical basis is most comprehensive considering all important interactions and possible intrusions in the molecule one may think of. The results are compared with earlier formulas which have been derived from more restricted theories. Finally, the method is applied to various examples demonstrating the use of the present procedure.

1 Introduction

Optical absorption and emission spectra can be classified according to the degree of resolution of their vibrational fine structure observed for electronic transitions. In general a distinction is made between narrow line and broad band spectra. The latter may also exhibit vibrational fine structure which is, however, much less resolved (with bands of generally about 100–300 cm^{-1} half widths or even more) than what is usually addressed as narrow line spectra (resolutions of some 5–20 cm^{-1} or less). Whether a spectrum is of narrow line or broad band type may be a matter of dispute and can be defined within other limits by other authors. From theory, narrow line spectra will be expected if during the transition no essential change of chemical bonding occurs. This is the case particularly for transitions between levels belonging to the same electron configuration (intra-configurational transitions). For complexes of transition metal ions, where the highest occupied orbitals are localized mainly on the metal ion, these transitions occur between levels belonging to a single electron configuration, i.e. t_2^n or e^n (in cubic notation), which is also the case for most complexes of lower (except square planar) symmetry since the cubic field is usually the main contribution to the ligand field. In principal, both narrow line and broad band spectra are observed from all electron configurations, except for d^1 and d^9 complexes for which in cubic symmetry only an inter-configurational transition, i.e. $t_2 \leftrightarrow e$, is possible (spin-orbit coupling being at the moment neglected). An exclusive observation of a narrow line spectrum for transitions from or into the respective ground states is not realistic since, in addition, transitions between different electron configurations which change general bond properties are always possible.

In practice, narrow line spectra mainly due to transitions within a *d*-electron configuration have been reported for d^2-(e.g. Ti(II), V(III) [1, 2]), d^3-(V(II), Cr(III), Mn(IV), Mo(III), Tc(IV), Re(IV) [3–14] and d^4-(Cr(II), Os(IV) [15–18]) ligand field configurations. For the numerous cases of inter-configurational transitions for which broad band spectra are expected, the vibrational fine structure is, in general, not resolved such that only very broad bands (1000–4000 cm^{-1} halfwidths) are observed for an electronic transition. However, in recent years an increasing number of spectra have been reported to show a more or less distinct fine structure which holds for transitions between levels belonging to different electron configurations, e.g. $t_2^n e^m \rightarrow t_2^{n-1} e^{m+1}$. This is achieved by choosing particular experimental conditions as, for instance, by measuring single crystals at very low temperature (below 4.2 K), using polarized light or, which turns out to be most efficient, by embedding the chromophore into a crystal of an optically inert host material (doping).

If the vibrational fine structure contains a series of peaks which occur at almost equal frequency intervals, i.e. if a progression is observed, the intensity distribution of these bands will allow a Franck-Condon analysis to be carried out. The progression is caused by a vibronic coupling which results from the interaction of electronic levels with a vibrational mode of particular symmetry [19]. Corresponding vibrational quanta are piled up in the spectrum forming the observed progression, the intervals of which can be used for identifying the type of coupling. The interpretation of these progressions will be the main topic of the present article. Progressions in absorption or emission spectra of transition metal compounds were reported from almost all *d*-electron configurations of central ions, i.e., from V(IV) [5, 20], Mo(V) [21], Os(VI) [22], Cr(III) [5, 23], Mn(II) [24], Ir(IV) [25, 26], Co(III) [27, 28], Rh(III) [29, 30], Pt(IV) [19, 30–32], Ni(II) [33, 34] and Cu(II) [35] where only the most common ions are mentioned. From higher transition group elements also binuclear complexes may exhibit extensive vibrational progressions which are in general due to metal-metal stretching vibrations [5, 36, 37]. In addition, closed shell complexes with s^2-configuration such as SeX_6^{2-} and TeX_6^{2-} (X = Cl, Br) should be mentioned which show well resolved progressions in the Jahn-Teller active mode of e_g symmetry (O_h) [38, 39]. Square planar Pd(II)- and Pt(II)-tetrachloro complexes are in so far an exception as these spectra exhibit extremely well resolved progressions although these spectra arise from inter-configurational transitions transferring an electron from lower mainly π-antibonding orbitals d_{xy}, d_{xz}, d_{yz} into strongly σ-antibonding $d_{x^2-y^2}$ orbital [40, 41].

In recent years vibrational progressions have been reported also for charge transfer transitions where electrons localized on ligands are excited into *d*-orbitals of the central metal (LMCT transitions) [5]. Examples of these cases are much less frequent. They have d^1-configurations (e.g. Mn(VI) [42]), d^2-(Mo(IV) [43]), d^3-(Re(IV) [44]), d^4-(Ru(IV), Os(IV) [45, 46] with relatively short progressions), d^5-(Ir(IV) [2, 5, 26, 47]) and d^8-*configuration* (Pd(II) [40], which has been interpreted also as allowed $d \rightarrow p$ transitions on the metal [48], Au(III) [49]). Vibrational progressions in the LMCT spectra of closed shell ions are well known from oxo- and thionato-metal complexes of type MA_4^{n-} with M = Cr(VI), Mn(VII) and Mo(VI) [50, 52]. Eventually also metal to ligand charge transfer (MLCT) transitions can be extremly well resolved; for instance, an extensive vibrational fine structure of Os(II) compounds (d^6) has been reported, which contrary to LMCT bands apparently does not show any lines which can be attributed to a band progression [53, 54].

In the theoretical part of this work we will calculate transition probabilities depending on possible shifts of potential energy curves due to vibronic coupling, which give rise to vibrational progressions in the electronic spectra. Comparing the calculated intensities with the experiment we shall present some specific examples demonstrating the usefulness of the theoretical model. The quantitative aspect of the procedure is dealt with by fitting the theoretical band shape function to experimental profiles of the spectra which allows the determination of the model parameters in the semiempirical model. From these, the geometry, potential energy curves and spectroscopic constants of the excited state are calculated which due to the short life time cannot be obtained otherwise.

2 Theory

2.1 The General Band Shape Function

Since vibronic coupling effects are in general small compared to energy differences of neighboring electronic levels one can apply perturbation theory when these interactions are to be considered. Only under this condition which is the adiabatic approximation the calculations will have physical significance [55]. In this framework the Herzberg-Teller coupling scheme is applicable [56] which considers the full Hamiltonian H(q, Q) of interactions between all electrons and nuclei by expansion into a Taylor series at fixed unclear coordinates Q_i^0 ($i = 1, 2, \ldots, 3N-6$), the zero-order term $H(q, Q^0)$ being the reference Hamiltonian of the corresponding pure electronic Schrödinger equation

$$H(q, Q) = H(q, Q^0) + \sum_{i=1}^{3N-6} \left(\frac{\partial H}{\partial Q_i}\right)_{Q^0} (Q_i - Q_i^0) + \frac{1}{2} \sum_{i,j}^{3N-6} \left(\frac{\partial^2 H}{\partial Q_i \partial Q_j}\right)_{Q^0} (Q_i - Q_i^0)(Q_j - Q_j^0) + \cdots \quad (1)$$

Here q and Q symbolize the sets of all electronic and nuclear coordinates q_i ($i = 1, \ldots, 3n$) and Q_i ($i = 1, 2, \ldots, 3N-6$), respectively. The derivatives are taken at the coordinate values Q_i^0 and the summation runs over all nuclear coordinates of independent vibrations. The expansion may be carried out with respect to different types of nuclear coordinates, i.e. symmetry coordinates and normal coordinates of the ground or the excited states. If the Q_i's are normal coordinates and the Q_i^0's are taken at the potential minimum of an electronic state E_k the coordinate values are by definition $Q_i^0 = 0$ for all i. In this case the matrix elements of the electron dependent part in the second term of Eq. (1) should vanish due to the minimal condition, i.e.

$$\left\langle \Psi_k \middle| \left(\frac{\partial H}{\partial Q_i}\right)_0 \middle| \Psi_k \right\rangle = 0 \quad \text{for all } i \quad (2)$$

where $\Psi_k(q)$ is an eigenfunction of the electronic Hamiltonian $H(q, Q^0 = 0)$ which is the first operator term in Eq. (1). If the calculation is carried out from an approximate eigenfunction, which in an ab initio calculation is usually the case (since $H(q, Q^0)\Psi_k(q) = E_k\Psi_k(q)$ cannot be solved accurately), we define $Q^0 = 0$ such that Eq. (2) is fullfilled, i.e., the origin of the normal coordinate is set at the calculated (approximate) potential minimum. In case E_k is the ground state, the coordinates $Q_i^0 = 0$ fixed at the origin are identical with the equilibrium coordinates of the nuclear framework.

First order terms in Eq. (1) due to vibronic coupling may in general give rise to changes of the electronic wavefunctions. It can be easily seen that eigenfunctions $\Psi_k(q)$ of the zero-order Hamiltonian $H(q, 0)$ may be intermixed by first order perturbation yielding

$$\Psi'_k(q, Q) = \Psi_k(q) + \sum_{l \neq k} c_{lk}\Psi_l(q) \tag{3}$$

where coefficients with non-diagonal elements

$$c_{lk} = \frac{1}{E_l - E_k}\left\langle \Psi_l \left| \sum_i \left(\frac{\partial H}{\partial Q_i}\right)_0 \cdot Q_i \right| \Psi_k \right\rangle \tag{4}$$

are in general finite leading to a wavefunction Ψ'_k which now is also a function of nuclear coordinates Q_i. Since this intermixing is different for each electronic state k, we can expect that potential energy curves due to vibronic coupling will be different for each electronic level k. One can see, however, that potential energy curves of the level k will not be changed (or shifted) by linear vibronic coupling if the normal coordinates refer to this particular state k. This can be checked by calculating the matrix element

$$\left\langle \Psi'_{k=0}(q, Q) \left| H(q, 0) + \sum_i \left(\frac{\partial H}{\partial Q_i}\right)_0 \right| \Psi'_{k=0}(q, Q) \right\rangle$$

with the wavefunction of Eq. (3) and considering the condition in Eq. (2), since all $\Psi_l(q)$ are eigenfunctions of $H(q, 0)$, i.e.

$$\langle \Psi_k(q) | H(q, 0) | \Psi_l(q) \rangle = 0 \quad \text{for } k \neq l$$

If k refers to the ground state, mixing of higher wavefunctions into the ground state is therefore not possible in first order and can be only due to terms of higher order in Q_i. In the case when the nuclear coordinates Q_i are those of the ground state, linear vibronic coupling is only possible in excited states leading to potential energy minima which compared to the ground state, are shifted along those coordinates Q_i for which the mixing coefficients Eq. (4) are finite. The so-called restoring forces of vibration of stability forces which after expansion drive the expanded system back to equilibrium are contained in the third term of Eq. (1) the quadratic part of which is

$$\frac{1}{2}\sum_i \left(\frac{\partial^2 H}{\partial Q_i^2}\right)_{Q^0} Q_i^2$$

This is identical to the potential energy operator of the harmonic oscillator. The excited state potential shift along Q_i depends on the integral [19]

$$A_i^e = \left\langle \Psi_e(q) \left| \left(\frac{\partial H}{\partial Q_i} \right)_0 \right| \Psi_e(q) \right\rangle \tag{5}$$

where $\Psi_e(q)$ is the electronic wavefunction of the excited state. A_i^e is the linear coupling parameter of this level coupled to the vibration along the coordinate Q_i. Assuming harmonic force fields the potential minimum is moved with respect to the ground state curve by

$$\Delta Q_i^e = -\frac{A_i^e}{k_i^e} \tag{6}$$

and the energy is stabilized by

$$\Delta E_i^e = -\frac{1}{2}\frac{(A_i^e)^2}{k_i^e} \tag{7}$$

These equations are obtained from the energy expression $E_i^e = k_i^e Q_i^2/2$ of a harmonic oscillator when shifting the parabola in the E_i^e, Q_i plane [19][1]. The force constant k_i^e belonging to this state is connected to one of the quadratic contributions in the third term of Eq. (1) by

$$k_i^e = \left\langle \Psi_e(q) \left| \left(\frac{\partial^2 H}{\partial Q_i^2} \right)_0 \right| \Psi_e(q) \right\rangle \tag{8}$$

The shift ΔQ_i^e of the curve minimum gives rise either to larger or smaller bond distances depending on the sign of A_i^e in Eq. (6). The coupling constants A_i^e associated to the energy level E_i^e can be expressed by orbital coupling coefficients when $\Psi_e(q)$ is written in antisymmetric products of single electron functions [19]. The stabilization energy ΔE_i^e of Eq. (7) depends only on the absolute value of A_i^e and is identical for either an increase or decrease of bond distances. The degree of vibronic coupling given by the coupling constant A_i^e determines the shape (band profile) and in particular the length of a progression to be expected in absorption and emission spectra. When a long vibrational progression is observed in an electronic transition we can conclude that strong vibronic coupling is present, for short progressions small vibronic coupling effects are expected. Also if no vibrational fine structure can be resolved, broad bands are indicative for strong vibronic coupling and narrow bands for weak vibronic coupling.

For a quantitative treatment of establishing connections between vibronic coupling and vibrational progressions in electronic spectra, band profiles from vibronic wavefunctions must be calculated using common procedures of time-dependent perturbation theory and Fermi's golden rule [57]. For emission, e.g., the transition rate which is the transition probability per unit time summed over

[1] Equations (6) and (7) refer to vibronic couplings with totally symmetric vibrations. For other couplings see Ref. [19]. Different formulas in literature refer to alternate definitions of A_i^e.

all degenerate components j and k of the excited (e) and ground state (g), respectively, is calculated by

$$W_{eg}(\nu) = \frac{4\nu^3}{3\hbar c^3} \sum_{j,k} \sum_{n,m} p(\mathrm{T}; m) |\langle \Psi_{ej}(\mathrm{q}) \Phi^e_m(\mathrm{Q}) | \mathrm{M} | \Psi_{gk}(\mathrm{q}) \Phi^g_n(\mathrm{Q}) \rangle|^2 \cdot \frac{2\vartheta}{(\mathrm{E}^e_m - \mathrm{E}^g_n - h\nu)^2 + \vartheta^2} \tag{9}$$

Here, instead of the usual delta function $\partial(\mathrm{E}^e_m - \mathrm{E}^g_n - h\nu)$, which supplies only discontinuous spectral lines, a Lorentz function with a line width 2ϑ (normalized to one) is introduced for all vibrational levels labeled by n and m of electronic states j and k which contribute to the transition rate of Eq. (9) [58]. M is the transition operator, in general, pertinent to an electric dipole transition. The nuclear wavefunctions are approximated by products of harmonic oscillator functions

$$\Phi^e_m(\mathrm{Q}) = \prod_i^{3N-6} \varphi^e_{m_i}(\mathrm{Q}_i) \quad \text{and} \quad \Phi^g_n(\mathrm{Q}) = \prod_i^{3N-6} \varphi^g_{n_i}(\mathrm{Q}_i) \tag{10}$$

where the $\varphi^e_{m_i}$ and $\varphi^g_{n_i}$ have in general displaced origins and vibrational frequencies ω^e_i and ω^g_i different in the excited and the ground state. The electronic wavefunctions $\Psi(\mathrm{q})$ must be prepared in such a way that the integral in Eq. (9) is finite, i.e. if Ψ_e and Ψ_g belong to electronic states, for which the transition is forbidden by symmetry or spin selection rules, wavefunctions of appropriate symmetry must be mixed into Ψ_e and Ψ_g by virtue of vibronic or spin-orbit coupling making the transition partially allowed [58, 59]. This is equivalent to breaking symmetry selection rules by considering additional interactions. The temperature dependent part in Eq. (9)

$$p(\mathrm{T}; m) = \prod_i 2\sinh\left(\frac{\hbar\omega^e_i}{2\mathrm{kT}}\right) \exp\left[-\left(m_i + \frac{1}{2}\right)\frac{\hbar\omega^e_i}{\mathrm{kT}}\right] \tag{11}$$

is due to the Boltzmann distribution over the vibrational levels of the initial state which in case of emission is the excited electronic state.

For evaluating Eq. (9) one rewrites the Lorentz function as a Fourier integral

$$\frac{2\vartheta}{(\mathrm{E}^e_m - \mathrm{E}^g_n - h\nu)^2 + \vartheta^2} = \frac{1}{\hbar} \int_{-\infty}^{\infty} dt \exp[it\hbar^{-1}(2\pi\hbar\nu + \mathrm{E}^g_n - \mathrm{E}^e_m) - \vartheta\hbar^{-1}|t|] \tag{12}$$

which separately holds in the harmonic oscillator approximation also for all energy components $\mathrm{E}^g_{n_i}$ of E^g_n and $\mathrm{E}^e_{m_i}$ with respect to each component of vibrational freedom. In this case we have

$$\mathrm{E}^g_{n_i} = (n_i + \tfrac{1}{2})\hbar\omega^g_i \quad \text{and} \quad \mathrm{E}^e_{m_i} = (m_i + \tfrac{1}{2})\hbar\omega^e_i + \hbar\Omega_i \tag{13}$$

where $\hbar\Omega_i$ is the electronic energy gap between the excited state and the ground state minima of potential energy curves in the plane defined by the energy axis and the coordinate Q_i.

With Eqs. (11) and (12), the transition rate, Eq. (9), for wavefunctions which are prepared so that the transition becomes allowed according to vibronic selection rules [58, 59], can be transformed into

$$W_{eg}(v) = \frac{4v^3}{3\hbar^2 c^3} \sum_i |F_i^{e-g}|^2 \int_{-\infty}^{\infty} dt\, K_i(t) \prod_j \exp[it(2\pi v - \Omega_j) - \vartheta\hbar^{-1}|t|]\, G_j(t) \tag{14}$$

where F_i^{e-g} is the transition moment which is finite due to vibronic mixing of electronic wavefunctions by virtue of the first order operator $(\partial H/\partial Q_i)_{Q^0}$ in Eq. (1). Vibrations labeled by the index i which give finite contributions to this intermixing, are named promoting modes, since they "promote" this transition in the sense that they make this transition possible by changing the vibrational state of the system when exciting the vibration $i \equiv p$ during the transition. The other modes $j \equiv a$ in the product of Eq. (14) are accepting modes since they serve as "acceptors" of the energy resulting from the transition in fulfillment of the energy conservation law. In the "single generating functions" $G_j(t)$, the Franck-Condon overlap factors [58, 60] are contained

$$G_j(t) = 2\sinh \frac{\hbar\omega_j^e}{2kT} \sum_{n_j, m_j} \exp\left[\left(n_j + \frac{1}{2}\right) it\omega_j^g - \left(m_j + \frac{1}{2}\right)\left(it\omega_j^e + \frac{\hbar\omega_j^e}{kT}\right)\right] \cdot |\langle \varphi_{n_j}^g | \varphi_{m_j}^e \rangle|^2 \tag{15}$$

and expressions for $K_i(t)$ are identical with $G_i(t)$ except for the last factor which contains the squared Herzberg-Teller matrix element

$$|\langle \varphi_{n_i}^g | Q_i | \varphi_{m_i}^e \rangle|^2$$

instead of the Franck-Condon factor. The $K_i(t)$ functions can be expressed by $G_i(t)$ functions yielding for the special case that normal coordinates and frequencies of promoting modes $i \equiv p$ is not changed by excitation $e \rightarrow g$, i.e., if $Q_p^e = Q_p^g \equiv Q_p$ and $\omega_p^e = \omega_p^g \equiv \omega_p$, (which is justified since promoting modes are in general angular vibrations leaving the chemical bond mainly unchanged) [58, 60]

$$K_p(t) = \frac{\hbar}{4M_p\omega_p} \left\{ \left[\coth\left(\frac{\hbar\omega_p}{2kT}\right) + 1\right] \exp(it\omega_p) + \left[\coth\left(\frac{\hbar\omega_p}{2kT}\right) - 1\right] \exp(-it\omega_p) \right\} G_p(t) \tag{16}$$

Introducing Eq. (16) into the transition rate formula of Eq. (14) we obtain

$$W_{eg}(v) = \frac{4v^3}{3c^3\hbar^2} \sum_p \left(\frac{\hbar}{4M_p\omega_p}\right) |F_p^{e-g}|^2 \cdot \left\{ \left[\coth\left(\frac{\hbar\omega_p}{2kT}\right) + 1\right] \prod_a \int_{-\infty}^{\infty} dt \exp[it(2\pi v - \Omega_a + \omega_p) - \vartheta\hbar^{-1}|t|] \cdot G_a(t) \right.$$

$$+\left[\coth\left(\frac{\hbar\omega_p}{2kT}\right)-1\right]\prod_a \int_{-\infty}^{\infty} dt \exp[it(2\pi\nu-\Omega_a-\omega_p)-\vartheta\hbar^{-1}|t|]\cdot G_a(t)\Bigg\} \tag{17}$$

The second line of Eq. (17) represents the Stokes term when transitional promotion is due to excitation of a promoting mode in the ground state (g), the third line is the anti-Stokes term originating from a promotion mode in the excited state (e).

The $G_j(t)$ functions of Eq. (15) have been calculated by Lin [60] when summing over Franck-Condon factors obtained from all possible (infinite) wavefunctions in the harmonic oscillator approximation. These $G_j(t)$ are rather complicated functions of the frequencies ω_j^g, ω_j^e and reduced masses M_j^g, M_j^e which are attributed to the corresponding normal coordinates Q_j^g and Q_j^e. They are collected in parameters describing the frequency relation β_j[2] and the potential minimum shift Δ_j of the excited state with respect to the ground state

$$\beta_j=\frac{\omega_j^e}{\omega_j^g} \qquad \Delta_j=\left(\frac{M_j^g\omega_j^g}{\hbar}\right)^{1/2}\Delta Q_j \tag{18}$$

Here ΔQ_j is the displacement of the potential curve minimum along the normal coordinate Q_j. For details concerning $G_j(t)$ one is referred to the literature [58, 60, 61, 62].

Moreover, the $G_j(t)$ functions also depend on the temperature, cf. Eq. (15), that has its origin from the Boltzmann factor Eq. (11). For simplification, we will now restrict them to low temperatures, i.e., to a region where vibrational modes of the excited state are essentially unoccupied at thermal equilibrium. From Eq. (11) we see that a vibrational level with a fundamental frequency of, e.g., $\omega_j = 200\ \mathrm{cm}^{-1}$, less than 1% will be populated at $T = 60$ K. In this case, no accepting modes $j \equiv a$ are occupied in the excited state since they usually correspond to stretching vibrations which have larger energy quanta. $G_j(t)$ then has a more convenient expression [59, 63]

$$G_j(t)=2\beta_j^{1/2}\exp\left[-it\delta_j-\frac{\Delta_j^2\beta_j(1-\exp(it\omega_j^g))}{1+\beta_j-(1-\beta_j)\exp(it\omega_j^g)}\right]$$
$$\cdot[(1+\beta_j)^2-(1-\beta_j)^2\exp(2it\omega_j^g)]^{-1/2} \tag{19}$$

in which $\delta_j=(\omega_j^e-\omega_j^g)/2$ corresponds to the difference of zero-point energies of the excited and the ground state.

From an inspection of Eq. (19) we see that we can remove the factor $\exp(-it\delta_j)$ from $G_j(t)$ and put it into the preceding exponential function of Eq.

[2] Since β_j is defined as frequency relation of the initial to the final state, it should be noticed that in absorption the reverse of β_j must be taken in Eq. (18).

(17). The integral in Eq. (17) then becomes

$$\int_{-\infty}^{\infty} \mathrm{d}t \exp[it(2\pi\nu - \Omega_a \pm \omega_p - \delta_a) - \vartheta\hbar^{-1}|t|]g_a(t) \tag{20}$$

with

$$G_a(t) = \exp(-it\delta_a)g_a(t) \tag{21}$$

Substituting $t = x_a/\omega_a^g$ and writing the integral representation of the exponential for the absolute value of x_a [64]

$$\exp\left\{\frac{\vartheta|x_a|}{\hbar\omega_a^g}\right\} = \frac{1}{\pi}\int_{-\infty}^{\infty} \mathrm{d}y \exp(-ix_a y)\frac{\vartheta/\hbar\omega_a^g}{y^2 + (\vartheta/\hbar\omega_a^g)^2} \tag{22}$$

we obtain from Eq. (20), since in $G_j(t)$ the variable t and x_a^g occur always as a product, i.e., $G_j(t) \equiv G_j(x_a)$,

$$\frac{1}{\pi\omega_a^g}\int_{-\infty}^{\infty'} \mathrm{d}y \int_{-\infty}^{\infty} \mathrm{d}x_a \exp\left\{ix_a\left[\frac{2\pi\nu - \Omega_a \pm \omega_p - \delta_a}{\omega_a^g} - y\right]\right\}$$

$$\cdot g_a(x_a)\frac{\vartheta/\hbar\omega_a^g}{y^2 + (\vartheta/\hbar\omega_a^g)^2} \tag{23}$$

Notice that the quotient in Eq. (23) is a Lorentz type function. Since $g_a(x_a)$ is a periodic function in 2π one can use for the integral over x_a the following relation

$$\int_{-\infty}^{\infty} \mathrm{d}x \exp(icx)g(x) = \sum_{n=-\infty}^{\infty} \int_{2\pi n}^{2\pi(n+1)} \mathrm{d}x \exp(icx)g(x)$$

$$= \sum_{n=-\infty}^{\infty} \exp(i2\pi nc)\int_0^{2\pi} \mathrm{d}x' \exp(icx')g(x') \tag{24}$$

where in the last part x is substituted by $x = x' + 2\pi n$ and c is a constant. Introducing the delta function $\partial(\ )$ with the identity [64]

$$\sum_{n=-\infty}^{\infty} \exp(2\pi ina) = \sum_{s=-\infty}^{\infty} \partial(s-a) \tag{25}$$

the integral in Eq. (23) will be

$$\frac{1}{\pi\omega_a}\int_{-\infty}^{\infty} \mathrm{d}y \sum_{n_a} \partial\left(\frac{2\pi\nu - \Omega_a \pm \omega_p - \delta_a}{\omega_a^g} + n_a - y\right)$$

$$\cdot\int_0^{2\pi} \mathrm{d}x_a \exp\left\{ix_a\left[\frac{2\pi\nu - \Omega_a \pm \omega_p - \delta_a}{\omega_a^g} - y\right]\right\}\cdot g_a(x_a)\frac{\vartheta/\hbar\omega_a^g}{y^2 + (\vartheta/\hbar\omega_a^g)^2} \tag{26}$$

For carrying out the integral over y we use the condition that due to the ∂-function this integral is finite only when

$$y = \frac{2\pi\nu - \Omega_a \pm \omega_p - \delta_a}{\omega_a^g} + n_a$$

by which Eq. (26) is transformed into

$$\frac{1}{\pi\omega_a}\sum_{n_a}\int_0^{2\pi} dx_a \exp(-in_a x_a) g_a(x_a) \frac{\vartheta/\hbar\omega_a^g}{[(2\pi\nu-\Omega_a\pm\omega_p-\delta_a)/\omega_a^g+n_a]^2+(\vartheta/\hbar\omega_a^g)^2} \tag{27}$$

which is simplified to

$$\frac{1}{\pi}\sum_{n_a}\int_0^{2\pi} dx_a \exp(-in_a x_a) g_a(x_a) \frac{\vartheta/\hbar}{(2\pi\nu-\Omega_a\pm\omega_p-\delta_a+n_a\omega_a^g)^2+(\vartheta/\hbar)^2} \tag{28}$$

With this the band profile (line shape) function Eq. (17) is

$$W_{eg}(\nu) = \frac{4\nu^3}{3\pi c^3\hbar^2}\sum_p\left(\frac{\hbar}{4M_p\omega_p}\right)|F_p^{e-g}|^2\left[\coth\left(\frac{\hbar\omega_p}{2kT}\right)+1\right]$$
$$\cdot\prod_a\sum_{n_a=0}^{\infty} \mathrm{I}(n_a;\Delta_a,\beta_a)\frac{\vartheta/\hbar}{(2\pi\nu-\Omega_a+\omega_p-\delta_a+n_a\omega_a^g)^2+(\vartheta/\hbar)^2} \tag{29}$$

where only the Stokes part of the spectrum is considered and in which

$$\mathrm{I}(n_a;\Delta_a,\beta_a) = \int_0^{2\pi} dx_a \exp(-in_a x_a) g_a(x_a) \tag{30}$$

is a distribution function which depends on the quantum number n_a of the vibrational mode a of the electronic ground state arising from the exponential factor in the integral and on Δ_a and β_a defined by Eq. (18), which enter by virtue of $g_a(x_a)$, i.e., by the single generating function $G_a(t)$ from Eq. (19). The sum over n_a in Eq. (29) now runs from zero to infinity since it can be shown that the distribution function $\mathrm{I}(n_a;\Delta_a,\beta_a)$ vanishes for negative n_a [59, 62].

In the denominator of Eq. (29) one can detect the "false origin" in the spectrum at the frequency $\Omega_a - \omega_p^g + \delta_a$ which, together with n_a quanta of the accepting mode ω_a^g, compensates the energy of incident light. The intensity maximum is at

$$2\pi\nu = \Omega_a - \omega_p^g + \delta_a - n_a\omega_a^g \tag{31}$$

the band width is $\vartheta/\hbar$.

The distribution function, Eq. (30), then describes the relative intensity of the vibrational members in the progression. If there is more than one accepting mode the corresponding progressions multiply their intensity which is a result of the product $\prod_a$ in Eq. (29).

2.2 Relation to Simplified Procedures

The relation of the preceding method to the common Franck-Condon analysis can be shown as follows. The integrand of Eq. (30) for any component a when introducing Eqs. (21) and (15), gives

$$g(x^g)\exp(-in'x^g) = 2\sinh\left(\frac{\hbar\omega^e}{2kT}\right)\sum_n\sum_m \exp\left[(n-n')ix^g - m\left(ix^e + \frac{\hbar\omega^e}{kT}\right) - \frac{\hbar\omega^e}{2kT}\right]|\langle\varphi_n^g|\varphi_m^e\rangle|^2 \tag{32}$$

Here it is $x^g = \omega^g t$ and $x^e = \omega^e t$ where the index a has been omitted. Combining the sinh with the last term $\hbar\omega^e/2kT$ in the bracket and restricting again to low temperature where no higher levels in the excited electronic state are occupied, i.e. $m = 0$, the righthand side of Eq. (32) becomes

$$\left[1 - \exp\left(-\frac{\hbar\omega^e}{kT}\right)\right]\sum_n \exp[(n-n')ix^g]|\langle\varphi_n^g|\varphi_m^e\rangle|^2 \tag{33}$$

With this, integration of Eq. (32) over x^g leads to

$$I(n'; \Delta, \beta) = \left[1 - \exp\left(\frac{-\hbar\omega^e}{kT}\right)\right]\sum_n \int_0^{2\pi} dx^g \exp[(n-n')ix^g]|\langle\varphi_n^g|\varphi_m^e\rangle|^2 \tag{34}$$

introducing again the distribution function of Eq. (30). For $T \rightarrow 0$ we finally obtain

$$I(n; \Delta, \beta) = 2\pi|\langle\varphi_n^g|\varphi_m^e\rangle|^2 \tag{35}$$

since the integrals are only finite for $n = n'$. We see that except for the factor 2π (which also can be incorporated into $I(n; \Delta, \beta)$ by choosing an appropriate definition in Eq. (30)) the distribution functions for $T \rightarrow 0$ are identical to the Franck-Condon factors. The usual analysis on the basis of Franck-Condon overlap factors is therefore only valid for the limit of low temperature; in this case any temperature dependence, also that caused by transition into promoting modes in the ground state, Eq. (29), is neglected. Formulas considering larger temperature dependences, where also accepting modes of electronically excited levels are occupied ($m \neq 0$), can be worked out and are given elsewhere [63], they will not be reported here. The theory has been experimentally tested on the luminescence spectrum of $[ReCl_6]^{2-}$ doped in K_2PtCl_6 type crystals measured at various temperatures [65].

A mathematical analysis of the distribution function in Eq. (30) is carried out without further approximation, it leads with the use of $G_a(t)$ of Eq. (21) and Eq. (19) [61, 62] to

$$I(n_a; \Delta_a, \beta_a) = \frac{2\beta_a^{1/2}}{1+\beta_a}\sum_{k=0}^{[n_a/2]}(-1)^k\binom{-1/2}{k}\left(\frac{1-\beta_a}{1+\beta_a}\right)^{2k}\tilde{I}(n_a - 2k; \Delta_a, \beta_a) \tag{36}$$

where

$$\tilde{\mathrm{I}}(n_a; \Delta_a, \beta_a) = \exp\left(-\frac{\beta_a \Delta_a^2}{1+\beta_a}\right)\left(\frac{1-\beta_a}{1+\beta_a}\right)^{n_a} \sum_{k=1}^{n_a} \frac{1}{k!}\binom{n_a-1}{n_a-k}\left(\frac{2\beta_a^2 \Delta_a^2}{1-\beta_a^2}\right)^k \tag{37}$$

in which Δ_a, β_a are defined by Eq. (18) and the summation in Eq. (36) runs until $[n_a/2]$ which is the largest integer smaller or equal to $n_a/2$.[3] For $n_a = 0$, it is

$$\mathrm{I}(0; \Delta_a, \beta_a) = \exp\left(-\frac{\beta_a \Delta_a^2}{1+\beta_a}\right) \tag{38}$$

The equations acquire a simpler form if we substitute

$$b_a = \frac{1-\beta_a}{1+\beta_a} \quad \text{and} \quad c_a = \frac{\beta_a \Delta_a^2}{1+\beta_a} \tag{39}$$

by which Eq. (37) becomes [63, 66]

$$\tilde{\mathrm{I}}(n_a; b_a, c_a) = \exp(-c_a) \sum_{k=1}^{n_a} \frac{1}{k!}\binom{n_a-1}{k-1} c_a^k (1-b_a)^k b_a^{n_a-k} \tag{40}$$

This is used in the line shape function of Eq. (29) and can be handled by computer programs more easily.

The extension of the formulas to degenerate accepting modes which occur when a Jahn-Teller effect in the excited state is present is relatively easy. In this case the products of distribution functions can be rewritten by convolution into fundamental distributions which does not change the overall expression of Eq. (29) [38, 66]. Also, it is possible to consider an intermixing of modes in the excited state by virtue of the bi-linear term in Eq. (1) (Duschinsky effect [67]). Since it is difficult to decide from most of the spectra if this effect is really observed in the case of the present complex compounds, we will not consider it here and refer to the literature [68, 69]. This is justified as long as we can explain the experimental spectra satisfactorily applying the "parallel mode" approximation leading to the line shape function of Eq. (29) as it has been described in the method above.

2.3 Comparison with Spectroscopic Band Profiles

For establishing the context to the experiment, we compare the line shape function of a transition, Eq. (29), with the corresponding band profile measured in absorption or emission at low temperature $\mathrm{T} \rightarrow 0$. The parameters Δ_a, β_a,

[3] For half integer α use $\binom{\alpha}{k} = \frac{\alpha(\alpha-1)\cdot\dots\cdot(\alpha-k+1)}{k!}$ valid for all real numbers of α and $\binom{\alpha}{0} = 1$.

ϑ and eventually $\Omega_a + \delta_a$ are determined by optimal fit (ω_a^g, M_p^g and M_a^g entering by Eq. (18) are known from the normal coordinate analysis carried out using infrared and/or Raman spectra). For higher temperatures, the fitting must be performed using a formula which considers appropriately the Boltzmann weighting factor $p(T;m)$ of Eq. (11). An inspection of the formulas show, however, that the relative intensities of the components in the band progression will not be changed within the approximation applied. Due to the smaller vibrational quanta of promoting modes, the temperature dependence resulting from the coth-factor in Eq. (16) at low temperature is larger than the temperature effects which are caused by occupations of accepting modes, $m_j \equiv 0$, in the excited state, cf. Eq. (15). For increased temperatures, the intrusion caused by the anti-Stokes term of Eq. (17) also becomes important. For more details the literature should be consulted [63, 65].

By evaluation of the parameters Δ_a, from Eq. (29), and ΔQ_a, from Eq. (18), we know the distortion of the excited state geometry and obtain the atomic equilibrium distances in relation to the ground state. This change is, as discussed previously, a consequence of vibronic coupling of this state to the accepting mode Γ_a. From fitting β_a to the band profile, we obtain the corresponding change of the vibrational quantum, cf. Eq. (18), and therefore of the potential energy curve valid at least close to the minimum. This allows, when the harmonic oscillator model is assumed again, the calculation of the corresponding force constant alteration due to excitation. Equations (6) and (7) then supply the coupling constant A_a^e and the stabilization energy ΔE_a^e which are attributed to Γ_a. Since only the square of ΔQ_a can be determined from the experiment by the fitting procedure (cf. Eq. (37)) the sign of ΔQ_a is unknown. Therefore a decision as to whether the molecule has larger or smaller atomic equilibrium distances in the excited state compared to the ground state cannot be made. Here, other methods must be brought forward, e.g., using arguments on the basis of bond strength changes. Expressing the normal coordinate ΔQ_a-values by vibrational elongations in cartesian coordinates Δx, Δy, Δz, the calculation of actual bond changes between each pair of atoms in the excited molecule is possible. In the case where more than one progression can be identified in the experimental spectrum (e.g., when evidently distortions due to accepting modes of a_{1g} and e_g symmetry become apparent in the spectrum [23, 27, 70]), the total bond changes are obtained by simply adding the distortions due to each contribution of distortions in the x, y and z directions.

Occasionally, measured band progressions are theoretically analyzed by applying the Poisson formula on the intensity distribution. According to this, the relative band intensity of the member n in the progression is

$$I^{(n)} = \exp(-\gamma)\frac{\gamma^n}{n!} \tag{41}$$

Here γ is a parameter which is a measure of the potential energy shift of the excited state relative to the ground state [23, 27, 70–72]. The adaptation of this distribution to the band maxima of each member in the progression supplies a numerical value for γ from which the desired coordinate shifts for the equilibrium distances are calculated.

One can easily see that the intramolecular distribution $\tilde{\mathrm{I}}$ of Eq. (40) adopts the Poisson distribution formula for $\beta_a = 1$, i.e., $b_a = 0$ due to Eq. (39), the sum containing only a single term $k = n_a$ since for all others the last factor in Eq. (40) vanishes for $b_a = 0$ [61, 62]. The resulting formula

$$\tilde{\mathrm{I}}(n_a; 0, c_a) = \exp(-c_a)\frac{c_a^n}{n!} \tag{42}$$

with $\beta_a = 1$, is identical to Eq. (41) for

$$\gamma = c_a = \frac{1}{2}\left(\frac{\omega_a^g M_a^g}{\hbar}\right)\Delta Q_a^2 \tag{43}$$

when using Eqs. (18) and (39)[4].

The authors who therefore apply the Poisson formula assume beforehand that $\beta_a = 1$, which means they assume equal frequencies (and force constants) for normal modes of the ground and excited state which is not always justified. Moreover, this formula can only be used for explaining intensities of a progression which are measured at low temperature, i.e., in the "zero-temperature limit' where the Boltzmann factor, Eq. (11), is neglected (set equal to 1). In the case of finite temperature, the integral intensity belonging to the transition is different, however the intensity distribution within a progression which is built up on a certain promoting mode p in Eq. (29), in our approximation, will not change with temperature.

For increased temperatures when vibrational levels of accepting modes are occupied, i.e., $m_i \neq 0$ in Eqs. (11) and (15), the distribution functions are distinctly more complicated leading to spectral progressions which differ from the low temperature expressions also qualitatively depending on how many quanta $m_i = 1, 2, \ldots$ are involved in the electronically excited state [63].

2.4 Excited State Time Evolution

Another way of analyzing band progressions, instead of starting with transition moments, results from considering the time development of the excited state. It proceeds from the time dependent wavefunctions, sets up the corresponding correlation function and obtains the energy distribution of the transition by Fourier transformation [73–75]. It can be shown that this analysis is completely equivalent to that of the common Franck-Condon procedure. We will briefly show that this correspondence holds when neglecting damping effects for simplicity [75].

---

[4] c_a is identical to the common Huang-Rhys factor S_a which occasionally is denoted also as vibronic coupling parameter. It is related to the true coupling parameter A_a^e of Eq. (5) by $S_a = \frac{\beta_a}{1+\beta_a}\frac{(A_a^e)^2}{\hbar M_a \omega_a^3}$.

If φ_n^g and φ_m^e are eigenfunctions of the Schrödinger equation of the vibrating nuclei for the electronic ground state (g) and excited state (e), the time development is given by the t-dependent Schrödinger equation $i\hbar\dot{\varphi} = \mathrm{H}\varphi$ with the general solution $\varphi(t) = \varphi \exp(-i\mathrm{H}t/\hbar)$ and, if H is independent of t and φ_n is an eigenfunction of H to the energy E_n, we have $\varphi_n(t) = \varphi_n \exp(-i\mathrm{E}_n t/\hbar)$. Following the evolution of the corresponding excited state function

$$\varphi_m^e(t) = \varphi_m^e \exp\left(\frac{-i\mathrm{H}t}{\hbar}\right) \tag{44}$$

we develop φ_m^e in terms of the complete set of ground state functions

$$\varphi_m^e = \sum_n c_{nm}\varphi_n^g = \sum_n \langle\varphi_n^g \mid \varphi_m^e\rangle\varphi_n^g \tag{45}$$

with the time development

$$\varphi_m^e(t) = \sum_n c_{nm}\varphi_n^g \exp\left(\frac{-i\mathrm{E}_n t}{\hbar}\right) \tag{46}$$

The time correlation function for the excited state m obtains with Eqs. (45) and (46)

$$P(t) = \langle\varphi_m^e \mid \varphi_m^e(t)\rangle = \sum_{n'}\sum_n c_{n'm}c_{nm}\left\langle \varphi_{n'}^g \mid \varphi_n^g \exp\left(\frac{-i\mathrm{E}_n t}{\hbar}\right)\right\rangle \tag{47}$$

and due to the orthogonality of the φ_n^g's this simplifies to

$$P(t) = \sum_n |c_{nm}|^2 \exp\left(\frac{-i\mathrm{E}_n t}{\hbar}\right) \tag{48}$$

Fourier transformation of $P(t)$ supplies the energy distribution

$$P(\mathrm{E}) = \frac{1}{2\pi}\int_{-\infty}^{\infty} P(t)\exp\left(\frac{i\mathrm{E}t}{\hbar}\right)\mathrm{d}t \tag{49}$$

which is written with Eq. (48)

$$P(\mathrm{E}) = \sum_n |c_{nm}|^2 \frac{1}{2\pi}\int_{-\infty}^{\infty} \exp\left[\frac{i(\mathrm{E}-\mathrm{E}_n)t}{\hbar}\right]\mathrm{d}t \tag{50}$$

With the integral representation of the delta function

$$\partial(x) = \frac{1}{2\pi}\int_{-\infty}^{\infty} e^{ikx}\,\mathrm{d}k \tag{51}$$

we get

$$P(\mathrm{E}) = \sum_n |c_{nm}|^2 \partial(\mathrm{E}-\mathrm{E}_n) \tag{52}$$

in which in view of Eq. (45) we obtain

$$|c_{nm}|^2 = |\langle\varphi_n^g \mid \varphi_m^e\rangle|^2 \tag{53}$$

which is the Franck-Condon factor if the φ-functions refer to one-dimensional oscillator functions belonging to a certain vibrational mode which corresponds to the accepting mode of the preceding section. If the wavefunctions represent the vibrational eigenstates of the total molecule, i.e., $\Phi(Q)$ of Eq. (10), then Eq. (53) contains the product of corresponding Franck-Condon factors. By comparison with Eq. (35) we obtain

$$|c_{n0}|^2 = \frac{1}{2\pi} I(n, \Delta, \beta) \tag{54}$$

The interpretation of band progressions by the time dependent procedure is therefore identical with the Franck-Condon analysis and, in the low temperature limit, to the method of molecular distributions as well. The line shape function obtained on the basis of Eq. (52) (for $E = h\nu$) differs under this condition from that of Eq. (12) only in the line shape function of each vibrational member in the progression which in Eq. (52) is the delta function and in Eq. (12) has a Lorentz type distribution.

In connection with the time-dependent picture of electronic transition a missing mode effect (MIME) has been postulated [75] trying to explain the vibrational progressions when they are measured in quanta which do not occur in the set of normal vibrational modes in the molecule. It has been shown that the total wave packet $\langle\Phi|\Phi(t)\rangle$ being a product of overlap factors, Eq. (47), of several displaced modes can lead, when Fourier transformed, to a spectrum with a progressional interval which is a mixture of the original normal modes. The spectrum of $W(CO)_5(py)$ on which this effect has been exemplified is, however, insufficiently resolved [75] to be used as a proof that the MIME in view of the uncertainty of the damping factor exists in reality.

3 Comparison with the Experiment

In this section we will apply the formulas on band profiles derived in the preceding section, to experimental spectra which are recorded in absorption or emission. For a direct comparison, which allows the determination of the parameter values of the theory, the bands originating from an electron transition must be obtained as well resolved as possible, unveiling the vibrational fine structure in such a way that at least relative intensities can be obtained. Various difficulties will arise if a background absorption due to other transitions is present. This must be removed by subtracting the spectral slope due to the neighbouring absorption. If the transition is forbidden by symmetry selection rules which generally is the case for *d–d* transitions of high symmetry complexes, the transition must be induced by virtue of vibronic coupling to promoting modes. Since this leads to superpositions of the vibrational progressions each only slightly shifted from the others, the spectral resolution into vibrational compounds must be of high quality and the spectrum possibly of narrow line type (see introduction). This requires a large effort carrying out the experiments and selective equipment to be used. Spectra of high resolution are obtained, e.g., from complexes doped in spectroscopically inert host lattices and from single crystals, both measured at low temperature (possibly lower than 2 K). Also applying polarized light helps to improve the quality of the spectra. For obtaining appropriate intensity distributions in the progression by attributing intensity data or extinction coefficients to each member of the progression, a band analysis and plotting curve derivatives have proved to be useful. This is achieved by decomposing the spectral progression into band components of Gaussian or Lorentz line shape functions. Most conveniently, the half widths ϑ of these band components are assumed to be equal. The data, which are to be compared with the theoretical band profiles, are the intervals of subsequent band peaks and the corresponding intensities given by absorption densities (also in arbitrary units) or if a quantitative analysis is possible by absorption coefficients.

In the following, we will present some cases of complex compounds the spectra of which exhibit vibrational progressions of a quality that can be used for carrying out a band analysis. These data then are worthwhile to be compared with the theoretical band profile functions as derived before.

3.1 Post-Transition Metal Complexes with s^2 Electron Configuration

Metal ions of main group elements with s^2 external electron configuration which are doped in various host lattices are subject to large spectroscopical and theoretical interest. This deals with mainly mono- and divalent ions as Ga^+, In^+, Te^+, Sn^{2+}, Pb^{2+} and possibly also anions in alkali halides AX (A = Na, K, Rb, Cs; X = Cl, Br, I) [76] or less numerous tri- and tetravalent ions as As^{3+}, Sb^{3+}, Bi^{3+}, Se^{4+} and Te^{4+} in appropriate host crystals, as elpasolites, K_2PtCl_6-type crystals and other compounds, in which these ions can be incorporated without changing the ionic charge on the central position of the metal substituted [39, 77, 78]. From higher charged s^2 ions the coordination compounds A_2MX_6 with A = alkali ions and X = Cl, Br belong to this group which we will treat in the following somewhat more in detail.

The absorption spectrum due to excitation into *sp* electron configuration shows in general three band features which are generally denoted by A, B and C ordered by increasing wavenumber [76–78]. They are assigned to the electronic levels $\Gamma_4^-(^3P_1)$, Γ_3^-, $\Gamma_5^-(^3P_2)$ and $\Gamma_4^-(^1P_1)$, respectively, which all arise from the *sp* electron configuration (cf. Fig. 1). The absorption and emission spectrum, for example of a $SbCl_6^{3-}$ complex, is illustrated in Fig. 2 [79]. The chloro- and bromo-complexes of Se^{4+} and Te^{4+} show similar spectral features [80]. The band profiles are, however, not due to vibrational progressions but can be attributed to Jahn-Teller effects in the excited state which are all orbital degenerate [76]. In absorption the spectra could not be further resolved even at

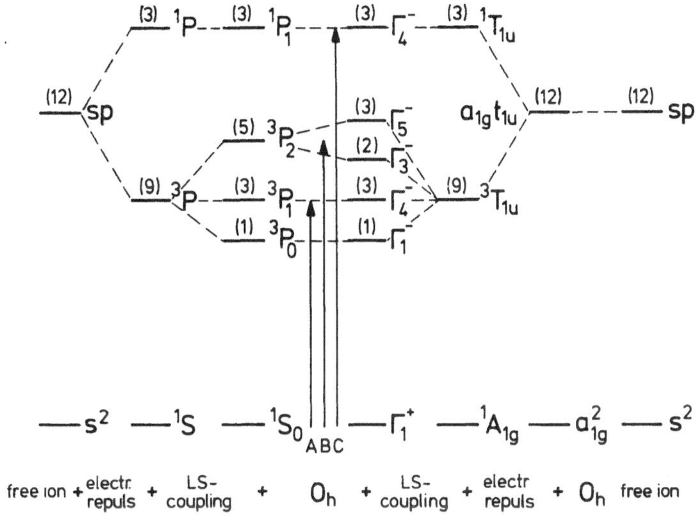

Fig. 1. Level scheme for $s^2 \rightarrow sp$ electron excitation considering various interactions, electron repulsion, octahedral O_h field and spin-orbit coupling ($L \cdot S$) plotted in different sequences

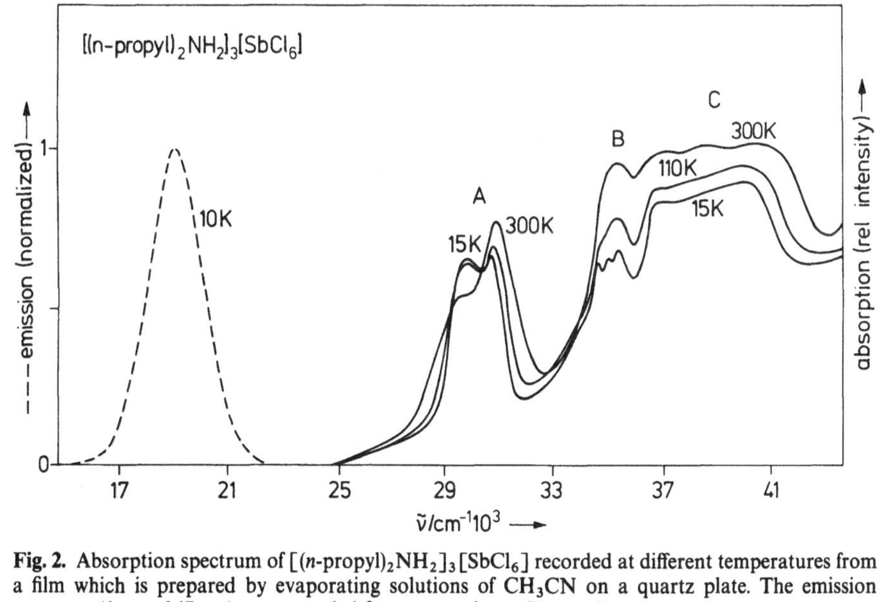

Fig. 2. Absorption spectrum of $[(n\text{-propyl})_2NH_2]_3[SbCl_6]$ recorded at different temperatures from a film which is prepared by evaporating solutions of CH_3CN on a quartz plate. The emission spectrum ($\lambda_{exc} = 367$ nm) was recorded from a powder at T = 10 K

very low temperature. In emission these complex ions show on the other hand very long well pronounced vibrational progressions with almost equal intervals up to high accumulations of quanta of a single vibrational mode. This can be obtained from powders of pure compounds [38, 82, 83] as well as from doped materials [39].

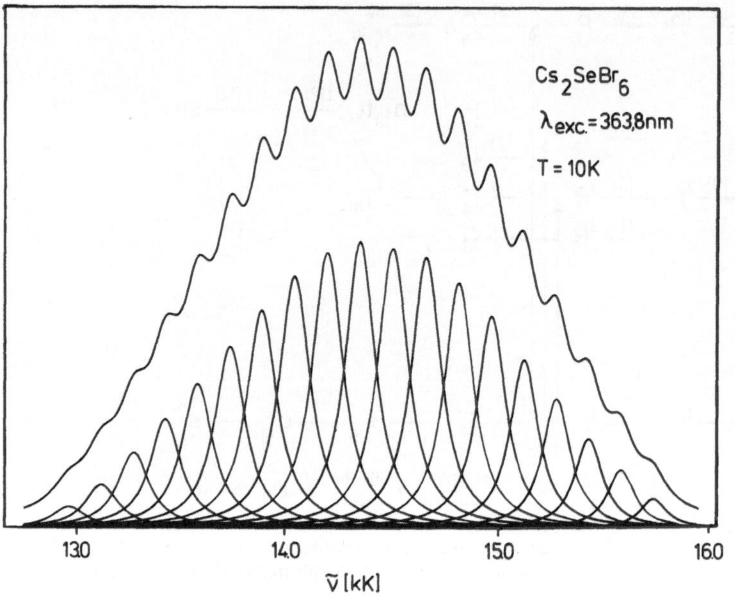

Fig. 3. Emission spectrum of Cs_2SeBr_6 (powder) at 10 K due to $\Gamma_4^-(^3P_1) \rightarrow \Gamma_1^+(^1S_0)$ transition obtained from argon laser excitation at 363.8 nm and band analysis by deconvolution into Lorentz curves

Figure 3 shows the emission spectrum of Cs_2SeBr_6 on excitation by laser light of wavelength $\lambda = 363.8$ nm at a temperature of about 10 K. The progression has intervals of $154\,cm^{-1}$ being equal up to high progressional members which in emission are found towards the low energy side (small wavenumbers) of the band. The half width of each component band is $2\vartheta = 170\,cm^{-1}$; as seen from band analysis, it increases by only $1\,cm^{-1}$ per member in the long wave region. Comparison with the vibrational spectra of Cs_2SeBr_6 indicates that the progressional interval is very close to the e_g vibrational fundamental which is measured in the room temperature Raman spectrum at $150\,cm^{-1}$ [80]. If emission occurs from the lowest symmetry allowed $\Gamma_4^-(^3P_1) \rightarrow \Gamma_1^+(^1S_0)$ transition by electric dipole radiation, we expect a Jahn-Teller effect in the excited state by relaxation to the absolute minimum on the potential hypersurface of this state. A calculation of this potential due to the Jahn-Teller distortion forces by the use of the Fukuda perturbation matrix of all *sp*-levels [84] yields two tetragonal minima in the $e_g(Q_2, Q_3)$ for each subspace level component *x*, *y*, *z* of Γ_4^- [85] (see Fig. 4). The excited state octahedron thus distorts due to Jahn-Teller forces while undergoing e_g vibrations leading to tetragonal symmetry followed by deactivation through radiative (and nonradiative) transition into e_g excited vibrational levels of the ground state. Note that due to $e_g \rightarrow a_{1g} + b_{1g}$ splitting when distorted to D_{4h}, the Franck-Condon factors attributed to the electronic transition in the a_{1g} space are finite. The spectrum

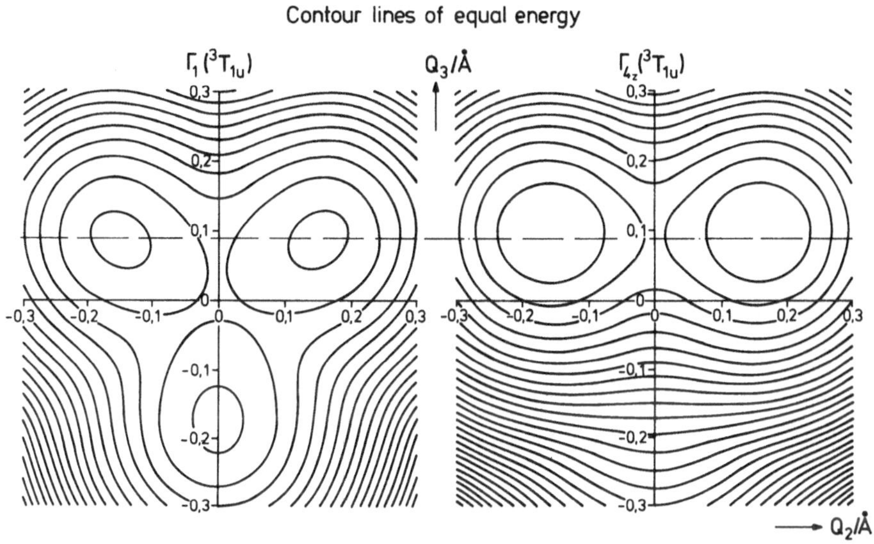

Fig. 4. Potential hypersurfaces of Γ_1^- ($^3P_0$) and Γ_4^- ($^3P_1$) levels in the $e_g(Q_2, Q_3)$ subspace calculated from the Fukuda coupling matrix with the vibronic coupling parameter $A_{e_g}^{\Gamma_4} = 4 \times 10^{-9}$ J/m from Eq. (5) (denoted by b in [84]), the electron (exchange) repulsion parameter G and spin-orbit parameter ζ are both set equal to 0.3 eV

which is produced by this mechanism is one depicted for instance in Fig. 3. We will see later that the kinetics might be much more complicated.

We will now use the formulas derived in the theoretical section for determining the molecular parameters of belonging to the excited Γ_4^- level, which are, e.g., the equilibrium distances, vibrational quanta, force constants, vibronic coupling parameters, Jahn-Teller stabilization, zero-phonon transition energies, etc. This is achieved by fitting the theoretical band profile function, Eq. (29), to the line shape of the experimentally measured emission spectrum. Since we obviously have only one accepting mode (the progression is in e_g quanta), the product $\prod_a$ of Eq. (29) has one factor $a = e_g$. As the transition $\Gamma_4^- \rightarrow \Gamma_1^+$ is allowed by electric dipole radiation, promoting modes are not necessary for activating this transition: the coth term and ω_p in Eq. (29) vanishes. The intensity distribution in the total band is given by the distribution function $I(n_{e_g}; \Delta_{e_g}, \beta_{e_g})$, the quotient in Eq. (29) determines the shape of each band component which is a Lorentz type curve with the line width 2ϑ. From this, the no-phonon transition $\omega_a^{0-0} = \Omega_a + \delta_a$ can be determined and, in case the electronic transition is forbidden, $\omega_a^{fo} = \Omega_a + \delta_a \pm \omega_p$ is the false origin in absorption (+) or emission (−) with respect to promotion by ω_p. Since in the present case, we have an allowed transition this does not apply. The distribution function is given by Eqs. (36) and (37) or more conveniently by Eq. (40), the latter function $I(n_{e_g}; b_{e_g}, c_{e_g})$ being dependent on b_{e_g} and c_{e_g} which are defined in Eq. (39). By comparison with the experimental band profile, the parameters b_{e_g} and c_{e_g} are adapted, e.g., applying a least square fit, one obtains [38]

$$b_{e_g} = -0.091 \quad \text{and} \quad c_{e_g} = 25.2$$

or, equivalently, by use of Eq. (39)

$$\beta_{e_g} = 1.2 \quad \text{and} \quad \Delta_{e_g} = 6.8$$

In general, normal modes of excited states are smaller than those in the ground state ($\beta < 1$) indicating flattened potential curves. In the present case, the potential curve is, however, squeezed to a smaller region in the e_g space due to the Jahn-Teller effect. With Eq. (18), the vibrational quantum of the accepting mode in the electronic excited state is

$$\omega^e_{e_g} = \beta_{e_g}\omega^g_{e_g} = 180\ \text{cm}^{-1}$$

and the alternation of equilibrium distance due to distortion from Eq. (18) is

$$\Delta Q_2(e_g) = 33\ \text{pm}$$

which corresponds to position changes of the ligands in cartesian coordinates

$$\Delta z = -2\Delta x = -2\Delta y = \frac{33}{\sqrt{3}} = 19\ \text{pm} \tag{55}$$

This is obtained from

$$Q_2(e_g) = \sqrt{\tfrac{1}{3}}(\Delta z_1 + \Delta z_2 - \tfrac{1}{2}\Delta x_3 - \tfrac{1}{2}\Delta x_4 - \tfrac{1}{2}\Delta y_5 - \tfrac{1}{2}\Delta y_6) \tag{56}$$

and

$$Q_3(e_g) = \tfrac{1}{2}(\Delta x_3 + \Delta x_4 - \Delta y_5 - \Delta y_6)$$

(notice that Q_2 and Q_3 are interchanged in [19] and [38]) making the tetragonal distortion with the main axis in z direction evident. From trigonal rotation of potential curve in the e_g space Q_2, Q_3 the equivalent distortion in x and y direction are calculated [38].

The coupling constant due to coupling of the excited $e = \Gamma_4^-(^3P_1)$ level to the $i = e_g$ vibrational mode is calculated by [19]

$$A^e_i = -\sqrt{3k^e_i}\Delta Q_i = -\sqrt{3M^e_i(\omega^e_i)^2}\Delta Q_i = -8.7\cdot 10^{-9}\ \text{J/m} \tag{57}$$

with the force constant $k^e_i = M^e_i(x^e_i)^2$. The Jahn-Teller stabilization energy is given by [19]

$$\Delta E^e_i = -\frac{1}{6}\frac{(A^e_i)^2}{k^e_i} = -\frac{1}{2}k^e_i(\Delta Q_i)^2 = -\frac{1}{2}\hbar\omega^e_i\Delta^2_i = -4160\ \text{cm}^{-1} \tag{58}$$

where Eqs. (18) and (57) are used, and the excited state wavenumber is $\hbar\omega^e_i = 180\ \text{cm}^{-1}$. Finally the zero-phonon energy (which is the zero-phonon transition into this state) is calculated from the band maximum $\hbar\omega_{max}$ of the progression by

$$\hbar\omega_{max} = \hbar\omega_{00} \pm n_{max}\hbar\omega_i \tag{59}$$

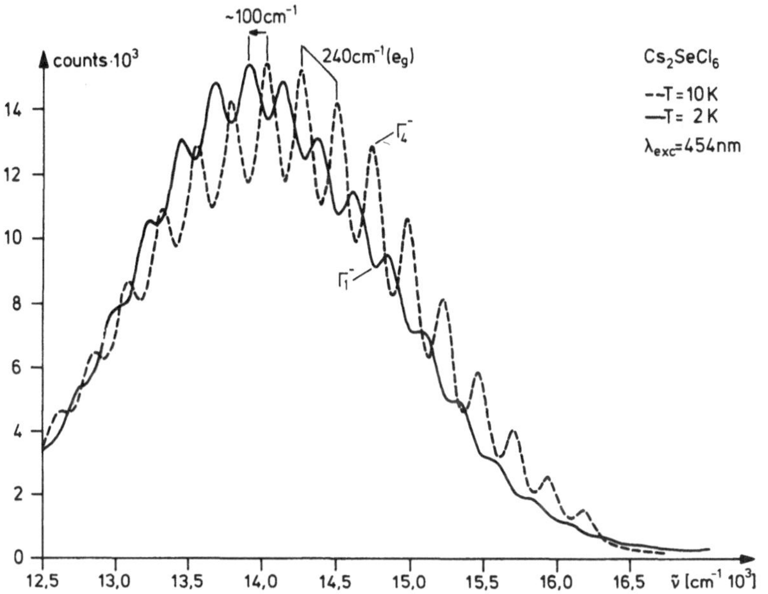

Fig. 5. Shift of Cs_2SeCl_6 emission spectrum by decreasing the temperature from T = 10 K to T = 1.9 K

(+ for absorption, – for emission) which in the present case becomes

$$\hbar\omega_{00} \equiv E_{00} = 17\,950\ \mathrm{cm}^{-1}$$

At even lower temperature (1.9 K) the spectrum of Cs_2SeCl_6 is observed to be shifted to lower energy by about 100 cm^{-1} (cf. Fig. 5) the progression being virtually unchanged. This spectrum has been attributed to emission from the Γ_1^- $(^3P_0)$ level which is somewhat lower in energy as Γ_4^- $(^3P_1)$ (cf. Fig. 1) from which emission at $T < 2.2$ K is uneffective due to lack of thermal occupation [85]. The progression in the e_g vibrational mode is explained by the hypersurface of the Γ_1^- $(^3P_0)$ level in the Q_2, Q_3 space of Fig. 4 which has three trigonal minima due to vibronic coupling to the higher Γ_3^- $(^3P_2)$ level (Fig. 1) by virtue of e_g vibrational modes (see the Fukuda perturbation matrix [84]). The minor quality of resolution in the progression at lower temperature also indicates that the emission at 1.9 K and 10 K results from different energy levels. The spectral shift of 100 cm^{-1} for the $SeCl_6^{2-}$ complex has been claimed not to be the energy difference of the electronic potential levels, this should rather be only 12 cm^{-1}, data resulting from kinetic measurements [78]. The difference to 100 cm^{-1} spectral shift is then explained by a t_{1g} vibrational mode which promotes the electronically forbidden $\Gamma_1^- \to \Gamma_1^+$ transition yielding an emission at $\omega = \omega(\Gamma_1^- - \Gamma_1^+) - \omega(t_{1g})$. The t_{1g} mode is, however, a cation–anion vibration attributed to a lattice mode where the framework of the $SeCl_6^{2-}$ complex remains unchanged (Fig. 6). Since this interpretation, obviously, has been confirmed by observing similar shifts of emission spectra when applying a magnetic

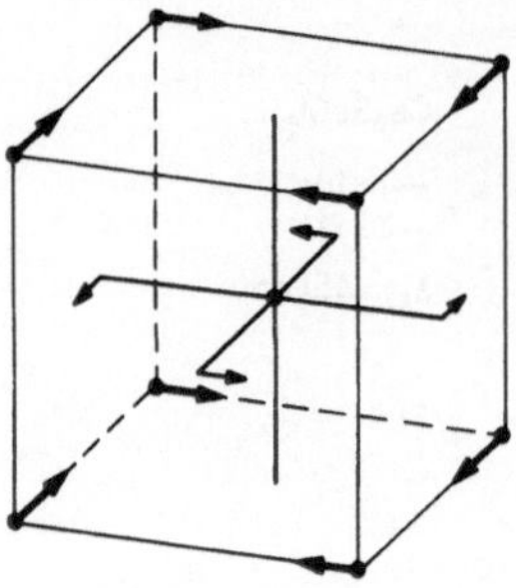

Fig. 6. Rotational cation-anion (lattice) vibration of t_{1g} symmetry

field which removes the necessity of vibrational promotion [86], we must assume that the *sp* electron levels of the complex can also be coupled to vibrations of atoms in the second coordination sphere. We know that electronic transitions and their vibrational side bands may very well be influenced by crystal fields of the counter ions. Here we evidently have the case that unique lattice vibrations can also serve as promoting modes for an electronic transition between levels localized primarily on the central metal of a coordination compound.

The situation is somewhat different for Cs_2TeCl_6. The emission spectrum at T = 10 K is shown in Fig. 7. The literature results reported on optical spectra and the kinetic investigations of this system agree quite well [38, 39, 81, 83]. According to the band analysis the progression also corresponds to e_g vibrational quanta of $\omega_{e_g} = 238\ \text{cm}^{-1}$. The parameters of the distribution function of Eqs. (36), (37) or (40) obtained by fitting the experimental spectrum to the band profile function Eq. (29) are in this case $\beta_{e_g} = 1$, $\Delta_{e_g} = 5.1$ and $b_{e_g} = 0$, $c_{e_g} = 13.0$, respectively[5]. From Eqs. (18), (56) to (58) the corresponding molecular constants are

nuclear distortion $\Delta Q_2(e_g) = 32$ pm $\quad \Delta z = -2\Delta x = -2\Delta y = 18$ pm

coupling constant $A_i^e = -6.3 \times 10^{-9}$ J/m

stabilization energy $\Delta E_i^e = -3100\ \text{cm}^{-1}$

zero-phonon energy $E_{00} = 19\,800\ \text{cm}^{-1}$

The calculated derivative spectrum in Fig. 7 shows a weaker second progression which obviously is due to emission from the lower Γ_1^- ($^3P_0$) state being symmetry forbidden.

A spectral shift of the emission spectrum towards lower temperature as reported for the Se compound has not been obtained. Probably an occupation of the lowest excited level Γ_1^- ($^3P_0$) is not essential because all radiation upon

[5] Notice that for $\beta_a = 1$ and $b_a = 0$, Eqs. (36), (37) and (40) are very simple, the sums contain only one term, i.e. for $k = 0$ or $k = n_a$, respectively.

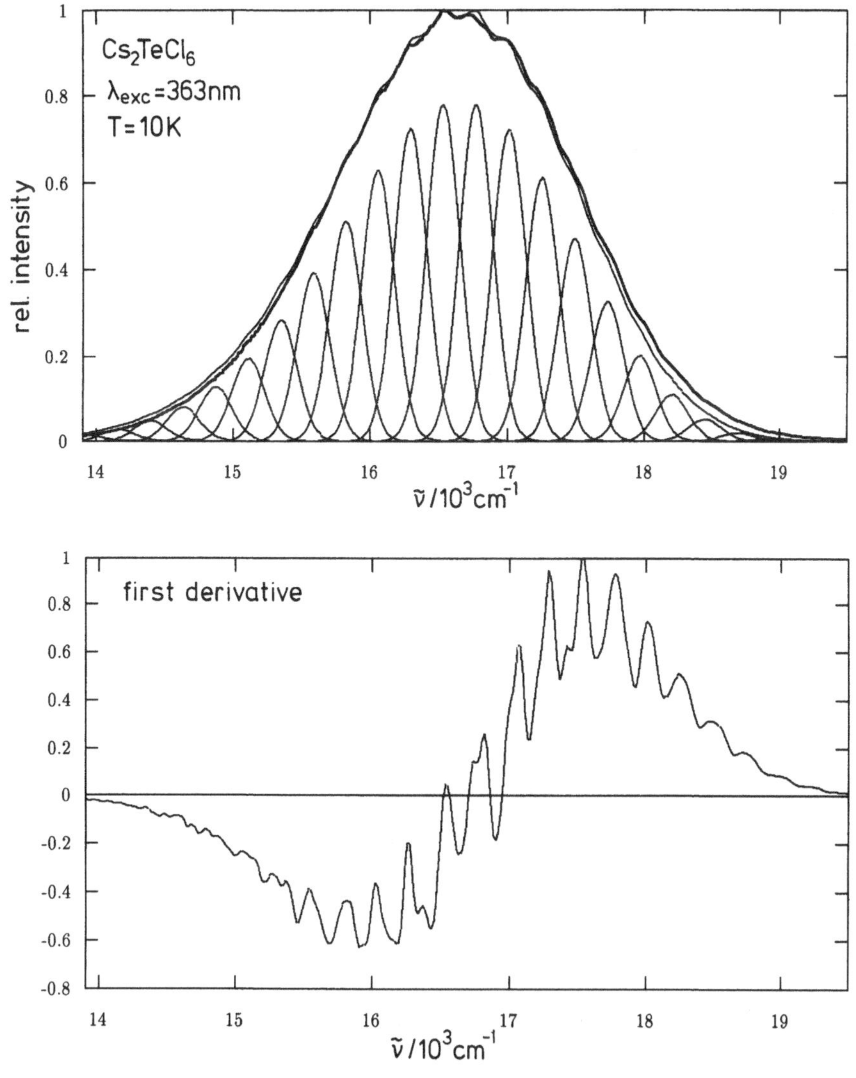

Fig. 7. Emission spectrum of Cs_2TeCl_6 (powder) at 10 K obtained from excitation by 363 nm (own measurement) and the calculated derivative spectrum ($dI/d\tilde{\nu}$ vs. $\tilde{\nu}$)

excitation at higher energy occurs from the $\Gamma_4^-(^3P_1)$ level [38, 39, 81] which is reached at an earlier stage during the decay process. Since the energy difference between $\Gamma_1^-(^3P_0)$ and $\Gamma_4^-(^3P_1)$ in $TeCl_6^{2-}$ is about 100 cm^{-1} [38, 39] the higher level cannot be occupied by thermal population at low temperature. If emission at this temperature only occurs from $\Gamma_4^-(^3P_1)$ the rate constant k_{20} for transition to the ground state $\Gamma_4^-(^3P_1) \rightarrow \Gamma_1^+(^1S_0)$ must be large compared to the rate constant k_{21}, which describes system distributions between levels $\Gamma_4^-(^3P_1)$ and $\Gamma_1^-(^3P_0)$ achieved predominantly by radiationless transitions.

A rate relation $k_{20} \gg k_{21}$ which is suggested by applying symmetry selection rules on the transition mechanics is referred to being present in the transition probability regime [87]. Earlier work on these compounds [39, 78, 81] has been based on assuming the reverse relation, i.e., $k_{21} \gg k_{20}$, anticipating that the thermal equilibrium between $\Gamma_4^-(^3P_1)$ and $\Gamma_1^-(^3P_0)$ is adjusting rapidly compared to emission. This situation corresponds to the thermal equilibrium regime. Recent luminescence decay measurements using excitation with different pulse lengths show, however, that the rate relation $k_{20} \gg k_{21}$, valid for the transition probability regime, is the appropriate one indicating that thermal equilibrium between the two excited states is not obtained [87]. This is particularly apparent for A_2SeCl_6 complexes with A = Rb and Cs [88].

Molecular parameters can be also calculated from absorption spectra if they are not resolved into vibrational fine structure. In this case, the continuous absorption profiles measured at different temperatures are compared with the theoretical line shape function which contains the sum of transition moments from the excited state eigenfunctions belonging to the *sp* electron configuration, weighted by the corresponding Boltzmann factors of the a_{1g}, e_g and t_{2g} vibrations in the ground state [89, 90]. Fitting the experimental band profiles of the Te^{4+} spectrum doped in K_2SnCl_6, taken at various temperatures between 13 K and 250 K, leads to vibronic coupling parameters of the $\Gamma_4^-(^3P_1)$ level coupled to e_g and t_{2g} vibrational modes [91]

$$A_{e_g}^{\Gamma_4^-} = -4.68 \times 10^{-9}\ \mathrm{J/m} \quad \text{and} \quad A_{t_{2g}}^{\Gamma_4^-} = -1.09 \times 10^{-9}\ \mathrm{J/m}$$

which are of similar magnitude. In contrast to emission, the coupling to all active vibrational modes must be considered for explaining band profiles in absorption (the a_{1g} coupling shifts all relevant levels by the same amount in first order and has not been evaluated in this calculation).

The e_g coupling parameter obtained from the absorption spectrum is, however, not fully comparable to that calculated from emission by Eq. (57) which for Cs_2TeCl_6 was $A_{e_g}^{\Gamma_4^-} = -6.3 \times 10^{-9}$ J/m. Inspection of Eq. (5) shows that for emission, the integral contains the electronic wavefunction calculated for nuclear coordinates at excited state equilibrium, in absorption this wavefunction must be taken at coordinates valid for the ground state equilibrium. Since, in the Oppenheimer approximation, the electronic wavefunction is, however, a slowly varying function of nuclear coordinates, we do not expect the coupling integral, Eq. (5), to change very much when taken at different equilibrium distances. The discrepancy may lie, for instance, in the shortcomings of the analysis of the absorption spectrum in which the vibrational quanta of the ground and excited states have been assumed to be equal ("effective fundamentals") [89, 90, 91], i.e., the frequency factor is set $\beta_i = \omega_i^g/\omega_i^e = 1$ for all i (cf. Eq. (18)). From emission we also have obtained for the e_g parameter $\beta_{e_g} = 1$, however, keeping in mind that, in the excited state, the bonding must be quite different, due to an occupation of a *p*- instead of an *s*-orbital on Te, the force constant must change to a certain amount which is not reflected in the β parameter value equal to

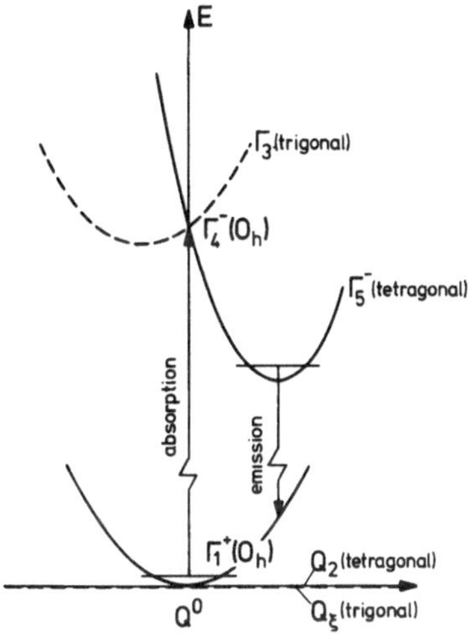

Fig. 8. Potential energy curves of $\Gamma_1^+(s^2)$ and $\Gamma_4^-(sp)$ (schematic) subject to Jahn-Teller distortions along $Q_2(e_g)$ (*full line*) and $Q_\xi(t_{2g}) = \sqrt{\frac{1}{3}}(Q_4 + Q_5 + Q_6)$ (*dotted line*) nuclear coordinates. Trigonal and tetragonal curves are depicted shifting towards different directions due to the opposite choice of signs of Q_2 and Q_ξ in the corresponding coupling matrices (cf. Ref. [19]). The other split levels resulting from the non-degenerate (Γ_2^-) are in either case higher in energy and are not illustrated for simplicity

unity. An inspecton of

$$A_i^e = -\sqrt{3}\omega(M\hbar\omega)^{1/2}\Delta \tag{60}$$

(derived from Eqs. (57) and (18)) shows that the coupling parameter A_i^e strongly depends on ω which for emission is the excited state frequency altering the coupling constant with respect to that of the ground state by a factor of $\beta^{3/2}$. A small error in adapting β will cause a larger error in A_i^e.

The discrepancy may be explained, as well, by the different systems investigated in emission (Cs_2TeCl_6) and absorption ($K_2SnCl_6{:}Te^{4+}$) or, more general, by the deficiencies of the theoretical models for calculating band profiles which are probably more serious for the analysis of the absorption spectrum. However, it can be concluded that the results obtained from different theoretical procedures, which are applied on different spectroscopical measurements, agree quite well in view of the approximations used.

With the two coupling parameters for e_g and t_{2g} modes obtained from absorption, the potential energy curves of the lowest level components of $\Gamma_4^-(^3P_1)$ are illustrated in Fig. 8. Distortion to tetragonal (in e_g space) or trigonal (t_{2g} space) symmetry is initiated by Jahn-Teller activity. It can be seen that emission occurs from the tetragonally relaxed potential sheet which has its equilibrium distance at lower energy. In absorption the final energy level is closer to the high symmetry $\Gamma_4^-(O_h)$ point where the trigonal and tetragonal potential curves are crossing. Due to the uncertainty in the nuclear position (in the excited state we also have a breakdown of Born-Oppenheimer approximation at this point [55]), each Franck-Condon transition arrives at different points on the excited energy curves, the absorption band profile reflecting the properties of the trigonal and tetragonal potential sheets.

3.2 The d^5 Ligand Field and Charge Transfer Transitions in Ir(IV) Complex Spectra

Octahedrally coordinated low spin d^5 complexes have particularly simple charge transfer spectra because electron transitions from ligand orbitals occur into the electron hole in the central metal t_{2g}^5 configuration filling up this electron shell. In this case only one multiplet level results for each orbital transition. Some of these bands have distinct vibrational structure, in particular well resolved band progressions which can be analysed [25, 26, 92]. We will deal with the results reported on $[IrCl_6]^{2-}$ and *trans*-$[IrCl_4F_2]^{2-}$ complexes for which several charge transfer bands have been observed as well as ligand field bands exhibiting distinct progression. The occurrence of resolved progressions for ligand field and charge transfer transitions in the same spectrum is a rare case and a comparison of corresponding parameters for the two transition mechanisms is certainly very interesting.

Figure 9 illustrates the 15 K absorption spectrum of $K_2SnCl_6:Ir^{4+}$ in the charge transfer region (A-region as denoted by Douglas [25]) arising from transitions from t_{2u} ligand orbitals to the empty t_{2g} orbital remaining in the t_{2g}^5 configuration of the Ir ion in the environment of O_h symmetry. Figure 10 depicts the 7 K ligand field spectrum of the same system in the ligand field region B which is much less intense [26]. A closer inspection of both figures

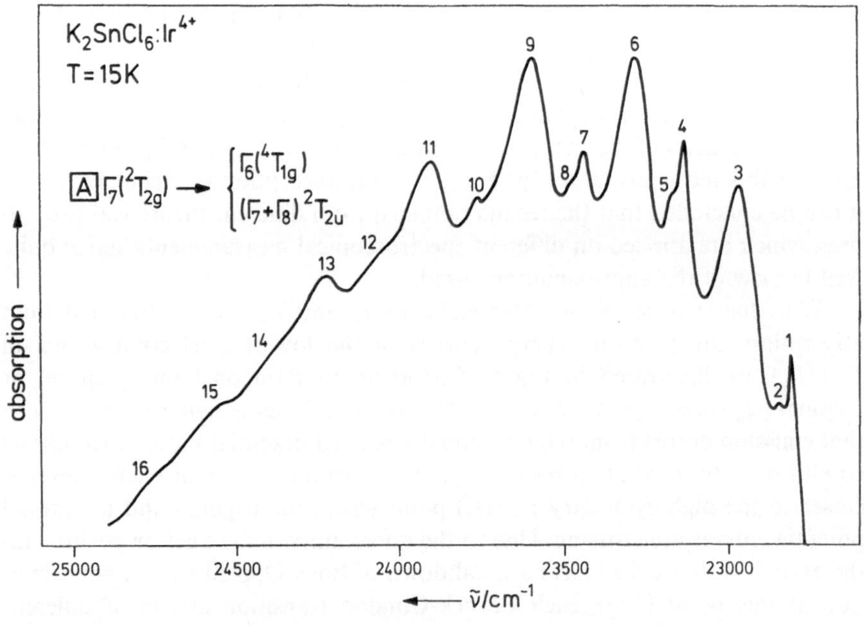

Fig. 9. Single crystal absorption spectrum of Ir^{4+} doped in K_2SnCl_6 in region A due to the charge transfer transition into $^2T_{2u}(t_{2u} \rightarrow t_{2g})$ at 15 K

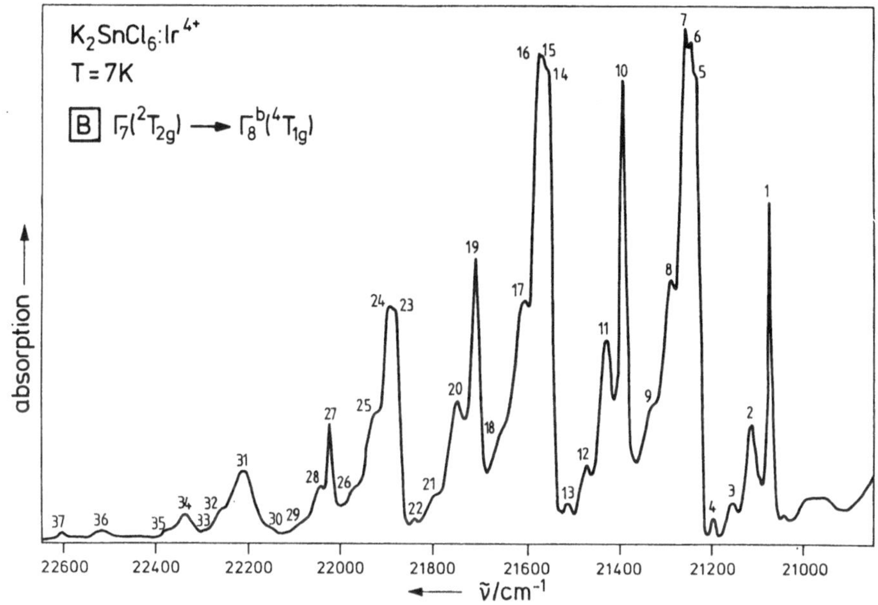

Fig. 10. High resolution single crystal absorption spectrum of Ir^{4+} doped in K_2SnCl_6 exhibiting progressions of ligand field transitions in the absorption gap between the first ($^2T_{1u}$) and second ($^2T_{2u}$) charge transfer bands (region B)

shows that several superimposed progressions of different band shapes but almost equal intervals of $\sim 310\ cm^{-1}$ (region A) and 315–320 cm^{-1} (region B) are present.

In Fig. 9 a progression of (1, 2), (4, 5), (7, 8) band pairs followed by less resolved numbers 10, 12, 14 and 16 can be detected, and another series 3, 6, 9, 11, 13, 15 of broader bands with higher (about nine times) oscillator strength. From comparison with the Raman spectrum one concludes that both progressions proceed in intervals of a_{1g} vibrational modes. The former series has been explained by a transition from the $\Gamma_7(^2T_{2g})$ ground state into the $\Gamma_6(^4T_{1g})$ ligand field level made allowed in O_h symmetry by $t_{2u}(\nu_6)$ and $t_{1u}(\nu_4)$ promoting modes which break the Laporte symmetry selection rule [26]. The second series is due to an electron transition from t_{2u} ligand orbitals into the empty t_{2g} metal orbital resulting in level transitions into the almost degenerate (due to the Ham effect [47, 93, 94]) Γ_7 and Γ_8 spin-orbit components of $^2T_{2u}$. The relative large intensity of the ligand field transition is attributed to an effective intensity borrowing mechanism from the Γ_7, $\Gamma_8(^2T_{2u})$ charge transfer level which is only a few 100 cm^{-1} higher in energy than the $\Gamma_6(^4T_{1g})$ ligand field level.

The vibronic band analysis of the progressions in the A region carried out as described in the theoretical section leads for the progressional series of the ligand field spectrum to the spectroscopic parameters

$$\beta_{a_{1g}} = 1.11 \quad \text{and} \quad \Delta_{a_{1g}} = 1.7$$

from which the molecular constants

nuclear distortion $\Delta Q_1(a_{1g}) = 9.5$ pm

and changes on the cartesian axes

$\Delta x = \Delta y = \Delta z = 3.9$ pm

coupling constant $A_{a_{1g}}^{\Gamma_6} = -1.9 \times 10^{-9}$ J/m

stabilization energy $\Delta E_{a_{1g}}^{\Gamma_6} = -450$ cm^{-1}

zero-phonon energy $E_{00} = 22\,800$ cm^{-1}

are derived. Fitting of the theoretical band profile function, Eqs. (29, 36, 37), to the experimental charge transfer series in this region yields the spectroscopic parameters

$\beta_{a_{1g}} = 1.08 \quad \Delta_{a_{1g}} = 2.0$

and the molecular constants

$\Delta Q_1(a_{1g}) = 11.1$ pm, i.e. $\Delta x = \Delta y = \Delta z = 4.5$ pm

$A_{a_{1g}}^{\Gamma_7,\Gamma_8} = -2.2 \times 10^{-9}$ J/m

$\Delta E_{a_{1g}}^{\Gamma_7,\Gamma_8} = -620$ cm^{-1}

$E_{00} = 22\,950$ cm^{-1}

From zero-point energies we can see that the electronic levels of the ligand field level and the charge transfer levels differ by only 300 cm^{-1} which favors strong vibronic coupling (possible by non-totally symmetric vibrations) giving rise to an effective intensity borrowing mechanism (see above). The smaller coupling constants (one order of magnitude decreased compared to those calculated for s^2-complexes) lead to smaller distortions ΔQ and lower stabilization energies ΔE.

In Fig. 10 depicting region B, several progressions of different intensity but comparable half widths are seen which are due to ligand field transitions [26]. Members of one progression occur at intervals of 315–320 cm^{-1} corresponding to the $\nu_1(a_{1g})$ stretching vibration of $IrCl_6^{2-}$. A triplet series (1, 2, 5/6), (10, 11, 14/15), (19, 20, 23), (27, 28, 31), and (34, 35, 37) is assigned to transitions into $\Gamma_8(^4T_{1g})$ ligand field level which are made allowed by $t_{2u}(\nu_6)$, $t_{1u}(\nu_4)$ and $t_{1u}(\nu_3)$, respectively, promoting modes in O_h notation. A second series (7, 8), (16, 17), (24, 25), (31, 32), (36) shifted about 180 cm^{-1} to higher energy is also due to $\Gamma_8(^4T_{1g})$. The presence of two series results from low symmetry field of the host lattice K_2SnCl_6 (tetragonal at low temperature) which split the (fourfold) Γ_8 level into two Kramers doublets. A ligand field calculation with appropriate parameters supports a D_{4h} splitting of Γ_8 by approximately 200 cm^{-1}. In addition, a third much weaker series (3, 4, 9), (12, 13, 18), (21, 22, 26), (29, 30, 33) can be detected differing from the first one by 80–85 cm^{-1} which we assigned to a lattice mode coupled to the lower ligand field component of $\Gamma_8(^4T_{1g})$ [26].

Remarkably, the combined electronic-lattice mode vibronic level $\Gamma_8 \times l$ obviously also needs a promotion of the complex by odd molecular vibrations in order to be detected in the *d–d* absorption spectrum.

From a fit to the theoretical band profile function Eqs. (29, 36, 37) to the triple progressional series one obtains for the spectroscopic parameters

$$\beta_{a_{1g}} = 1.08 \quad \Delta_{a_{1g}} = 1.5$$

which lead to the molecular constants

$$\Delta Q_1(a_{1g}) = 8.4 \text{ pm}, \quad \text{i.e. } \Delta x = \Delta y = \Delta z = 3.4 \text{ pm}$$

$$A_{a_{1g}}^{\Gamma_8} = -1.8 \times 10^{-9} \text{ J/m}$$

$$\Delta E_{a_{1g}}^{\Gamma_8} = -360 \text{ cm}^{-1}$$

$$E_{00} = 20\,900 \text{ cm}^{-1}$$

These parameters are quite similar to those derived for the spectrum in the A region (Fig. 9).

A further decrease of environmental symmetry is obtained if $IrCl_6^{2-}$ is doped in a lattice of lower symmetry. In Fig. 11, the absorption spectrum of this complex doped in $(Ethyl_3NH)_2[SnCl_6]$ crystals is illustrated in which Ir^{4+} is

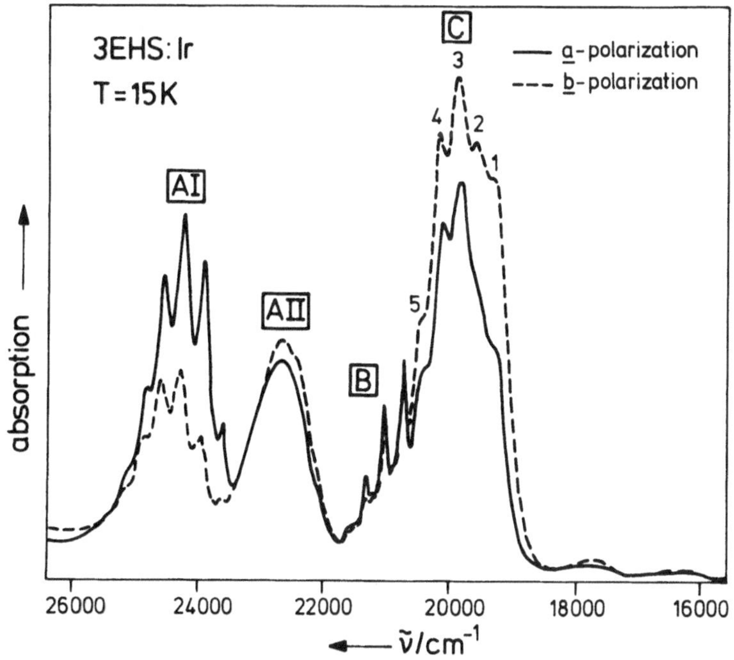

Fig. 11. Overall polarized absorption spectrum of 3EHS: Ir, i.e. Ir^{4+} doped in $(ethyl_3NH)_2[SnCl_6]$, at 15 K. Incident light perpendicular to the $(10\bar{1})$ crystal plane (see Ref. [26])

embedded in a lattice field of approximate D_{4h} symmetry. This field must be strong enough eliminating the Ham effect on $^2T_{2u}$ as observed for $K_2[SnCl_6]$ doped crystals since both components Γ_8 and Γ_7 are separately detected in absorption peaks A*I* and A*II*. The analysis of the main progression of peak A*I* in a_{1g} quanta of 305 cm^{-1} leads to the spectroscopic parameters

$$\beta_{a_{1g}} = 1 \text{ (assumed)} \quad \text{and} \quad \Delta_{a_{1g}} = 2.3$$

furnishing the molecular constants

$$\Delta Q_1(a_{1g}) = 13 \text{ pm}, \quad \text{i.e.} \Delta x = \Delta y = \Delta z = 5.3 \text{ pm}$$

$$A^{\Gamma_8}_{a_{1g}} = -2.5 \times 10^{-9} \text{ J/m}$$

$$\Delta E^{\Gamma_8}_{a_{1g}} = -810 \text{ cm}^{-1}$$

$$E_{00} = 23\,400 \text{ cm}^{-1}$$

A second progression, which is hardly to be detected in Fig. 11, proceeds in e_g quanta of 270 cm^{-1}, carrying out its analysis is much less certain because of the band profile which is not very distinct.

A well pronounced vibronic spectra is measured from $[(CH_3)_4N]_2[IrCl_4F_2]$ in which the Ir complex has a geometric *trans* configuration (cf. Fig. 12) [92, 95]. This spectrum is dominated by several charge transfer

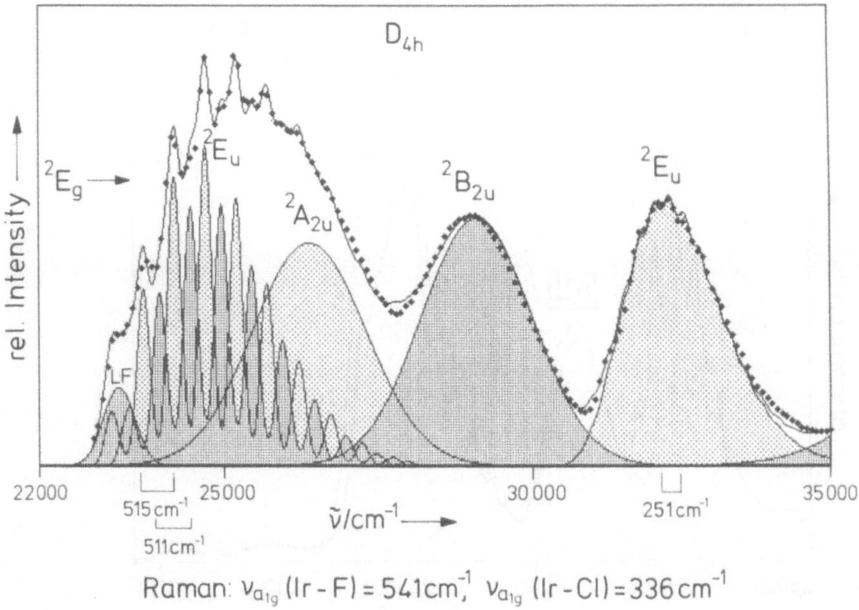

Fig. 12. Band analysis of the *trans*-$[IrCl_4F_2]^{2-}$ charge transfer absorption spectrum (10 K in KCl discs) by deconvolution into Gaussians (*points*: experimental, *solid line*: superposition of components)

bands resulting from the first two octahedral $^2T_{1u}$ and $^2T_{2u}$ charge transfer levels which are split by D_{4h} symmetry into ^2E_u, $^2A_{2u}$ and $^2B_{2u}$, ^2E_u, respectively. The transitions into ^2E_u levels exhibit vibrational fine structure which by band analysis can be identified resulting from superpositions of several progressions. These bands are vibrationally resolved because transitions from the ground state ^2E_g into ^2E_u are z-polarized shifting electron charge only into the z (F-Ir-F) direction. Since this change of electron-density does not cause rearrangements of nuclei other than those maintaining the symmetry of the molecule, only totally symmetric vibrations will be activated. Hence for transitions into ^2E_u only vibrations of this symmetry are invoked giving rise to progression in the $\nu_2(a_{1g})$ Ir-F stretching mode. The vibrational intervals 515 cm^{-1} and 511 cm^{-1} obtained from a band analysis of the two superimposed series due to spin-orbit components Γ_6^- and Γ_7^- of ^2E_u compare with the Raman wavenumber of 541 cm^{-1} for the $\nu_2(a_{1g})$ stretching mode [92] indicating decreased vibrational quanta in the excited state as it is usually the case. A fit of band profiles comparing the experimental absorption spectrum (Fig. 12) with the theoretical intensity function of Eqs. (29, 36, 37) leads to the spectroscopic parameter sets

(1) $\beta_{a_{1g}} = 1.04 \quad \Delta_{a_{1g}} = 2.9$

(2) $\beta_{a_{1g}} = 1.05 \quad \Delta_{a_{1g}} = 2.7$

which apply, respectively, to the ^2E_u components Γ_6^- and Γ_7^- which have their origin at $E_{00} = 23\,199$ cm^{-1} (1) and 23 477 cm^{-1} (2). A decision whether Γ_6^- or Γ_7^- is the higher (lower) spin-orbit level cannot be made. The molecular constants are calculated as

(1) $\Delta Q(a_{1g}) = 17$ pm

i.e., $\Delta z = 12$ pm for Ir–F bond length changes

$$\left.\begin{aligned} A_{a_{1g}}^{(1)} &= -5.1 \times 10^{-9}\ \text{J/m} \\ \Delta E_{a_{1g}}^{(1)} &= -2150\ \text{cm}^{-1} \end{aligned}\right\} \text{for (1)} = \Gamma_6^- \text{ or } \Gamma_7^-$$

(2) $\Delta Q(a_{1g}) = 15.9$ pm

i.e., $\Delta z = 11.2$ pm for Ir–F bond length changes

$$\left.\begin{aligned} A_{a_{1g}}^{(2)} &= -4.7 \times 10^{-9}\ \text{J/m} \\ \Delta E_{a_{1g}}^{(2)} &= -1850\ \text{cm}^{-1} \end{aligned}\right\} \begin{aligned} &\text{for (2)} = \Gamma_6^- \text{ or } \Gamma_7^- \\ &\text{alternatively to (1)} \end{aligned}$$

Although Γ_6^- and Γ_7^- result from the same parent state ^2E_u the parameters are rather different which is particularly surprising for the stabilization energy $\Delta E_{a_{1g}}$. The difference can be attributed to level intermixing to other parent states of the same symmetry, e.g., Γ_6^- and Γ_7^- of $^2T_{2u}$. Since a normal coordinate analysis is not available, the degree of mixing of a_{1g} metal ligand stretching modes ν_1(Ir-Cl) into ν_2(Ir–F) is not known. Therefore, the $\Delta Q(a_{1g})$ values are tentatively interpreted as deformations Δz of only Ir–F bonds. The numbers given for Δz are then in fact maximal values. The higher band arising from transition into ${}^2E_u$ resulting from ${}^2T_{2u}$ exhibits also pronounced vibrational structure. The band analysis yields a progression with similar vibrational quanta (501 cm^{-1}) which can be interpreted by the theoretical band profile formula as well. We do not want, however, to go into further detail on this, since spectroscopic and molecular parameters obtained for this transition are not much different from the preceding results.

The failure resolving the bands which result from transitions into ${}^2A_{2u}$ and ${}^2B_{2u}$ originates from their polarization (x, y) shifting charge between the x and y directions destroying the tetragonal symmetry of the molecule. For obtaining intensity also asymmetric stretching or angular modes in the x, y plane are invoked some of them having smaller vibrational quanta which cannot be resolved any more in a vibronic transition [92].

3.3 Ligand Field Transitions in Spectra of d^6 Complexes

Octahedral Co(III) compounds which are representatives of low spin d^6 complexes eventually have well resolved absorption and luminescence spectra with long vibrational progressions [27, 28, 72]. Some of them have been analysed by fitting the intensity distribution of the progressions to band profile functions obtained from different theoretical procedures [19, 27, 72]. At present we want to look at spectra of mixed ligand Co(III) complexes which exhibit well resolved vibrational progressions in their low energy spin-allowed transitions. Higher bands are less resolved due to superpositions of vibrational fine structure from lower transitions. In Fig. 13 the polarized absorption spectrum of *trans*-$[Co(NH_3)_4(CN)_2]Cl$ in the region of the first ligand field transitions into ${}^1A_{2g}$ and ${}^1E_g$ both resulting from ${}^1T_{1g}(O_h)$ is shown [96]. The long wavelength part can be recorded with higher sensitivity furnishing a largely resolved spectrum containing the pattern of fundamental bands (collection of false origins) which is repeated many times with varying intensities. In Fig. 14 it is demonstrated how this "fundamental spectral pattern" can be superimposed making up the experimental spectrum when an intensity distribution is underlaid to each progressional member. The progression proceeds in steps of 430 cm^{-1} which corresponds to the a_{1g} Co-NH_3 stretching vibration measured at 482 cm^{-1} in the 12 K Raman spectrum leading to a frequency effect of $\beta_{a_{1g}} = 482/430 = 1.12$. The band fit to the theoretical band profile curve supplies $\Delta_{a_{1g}} = 3.10$ from which the molecular parameters

$$\Delta Q(a_{1g}) = 20.6 \text{ pm}, \quad \text{i.e. } \Delta x = \Delta y = 10.3 \text{ pm}$$

$$A_{a_{1g}}^{{}^1A_{2g}} = -4.1 \times 10^{-9} \text{ J/m}$$

$$\Delta E_{a_{1g}} = -2050 \text{ cm}^{-1}$$

are calculated. A limited normal coordinate analysis considering only the two a_{1g} stretching modes of Co–NH_3 and Co–CN atomic groups shows that the 482 cm^{-1} quantum corresponds almost entirely to Co–NH_3. Another Raman

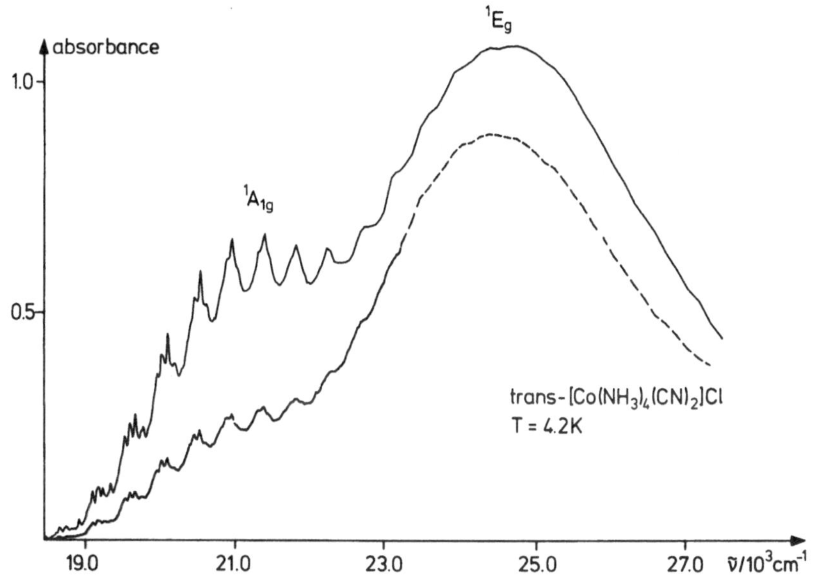

Fig. 13. Polarized absorption spectrum (low energy part) in two extinction directions (cf. Ref. [96]) of *trans*-$[Co(NH_3)_4(CN)_2]Cl$ at 4.2 K

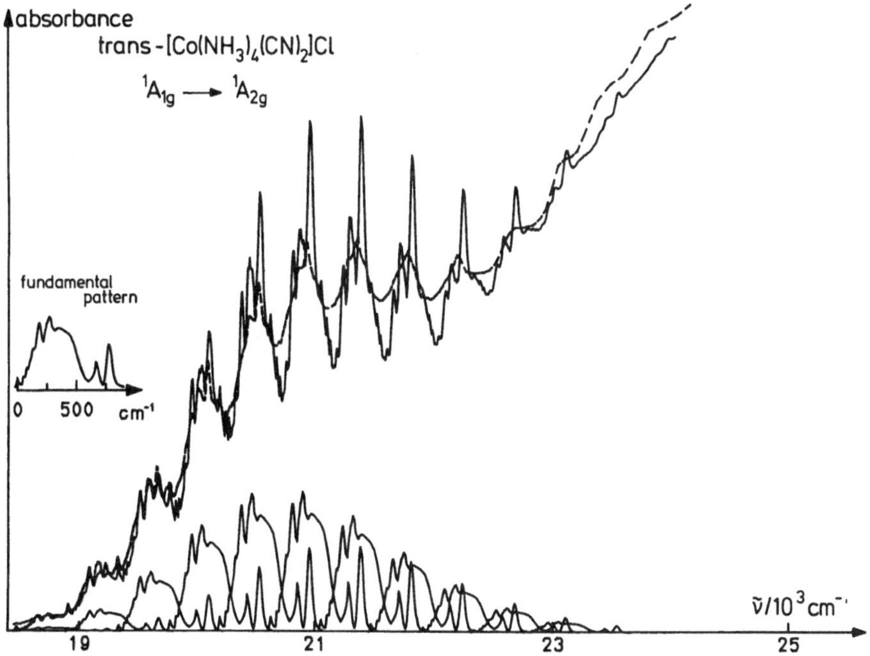

Fig. 14. Vibrational analysis of the $^1A_{1g} \rightarrow {}^1A_{2g}$ absorption band of the more intense component of the *trans*-$[Co(NH_3)_4(CN)_2]Cl$ polarized spectrum. *Dotted line*: experiment, *solid line*: superposition of propagated fundamental spectral pattern with (*upper part*) and without (*lower part*) considering the absorption tail from the higher transition into ^1E_g

band measured at 380 cm^{-1} belongs Co–CN vibrations. Therefore, in cartesian coordinates the distortions of the bonds in the x, y plane are calculated from the symmetry coordinate $\Delta Q(a_{1g}) = \frac{1}{2}(\Delta x_1 + \Delta x_2 + \Delta y_3 + \Delta y_4)$ which contains only Co–NH_3 contributions. The corresponding deuterated complex also exhibits a well resolved vibrational progression of the same transition. Although the fundamental spectral pattern and the vibrational interval of the progression are quite different, the molecular parameters are very similar to those of the protonated compound [96]. This, of course, is to be expected. The agreement can, however, be considered as a test for the accuracy of the results which are obtained from the present theoretical model.

Quite similarly proceeds the investigation of the optical spectrum of *trans*-$[Co(en)_2(CN)_2]ClO_4$ (en = ethylenediamine) which is illustrated in Fig. 15 [97]. The low energy band exhibiting vibrational structure is due to $^1A_g \rightarrow {}^1B_g$ transition (C_{2h} molecular point symmetry), the higher is assigned to $^1A_g \rightarrow {}^1A_g$, ^1B_g which are the other components of octahedral $^1T_{1g}$ when reducing the symmetry. Figure 16 depicts the result of a band analysis which is carried out by a repeated superposition of the fundamental spectral pattern in steps of 236 cm^{-1} with growing intensity towards higher energy. The underlaid intensity distribution when compared to the theoretical band profile function, Eqs. (29, 36, 37), supplies the spectroscopic parameters $\Delta_{a_g} = 4.8$ and $\beta_{a_g} = 1.15$ which refers to an a_g deformation mode of 269 cm^{-1} as obtained from the infrared spectrum [97]. For calculating the molecular constants of the *trans*-$[Co(en)_2(CN)_2]^+$ complex a normal coordinate analysis is necessary which has

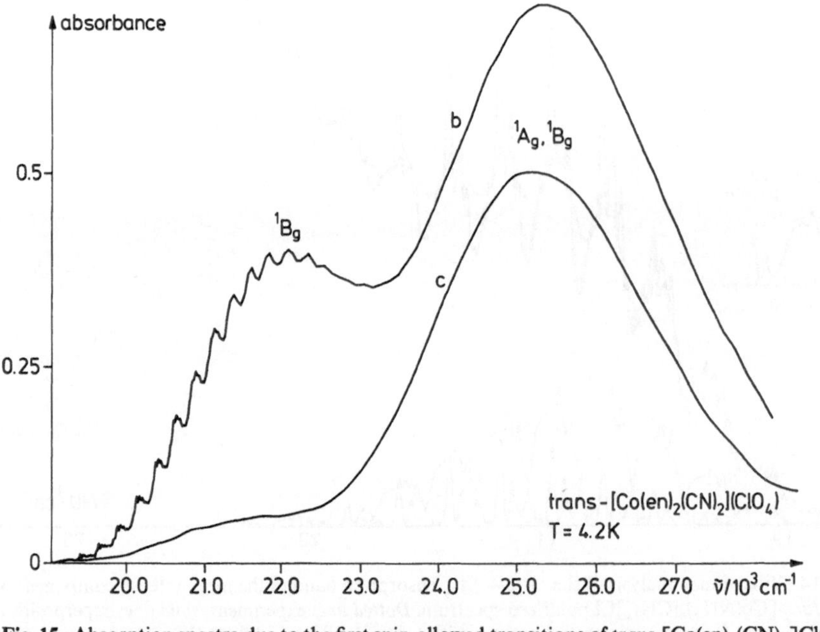

Fig. 15. Absorption spectra due to the first spin-allowed transitions of *trans*-$[Co(en)_2(CN)_2]ClO_4$ at 4.2 K polarized parallel to the b and c crystal axes recorded from plates of different thickness [97]

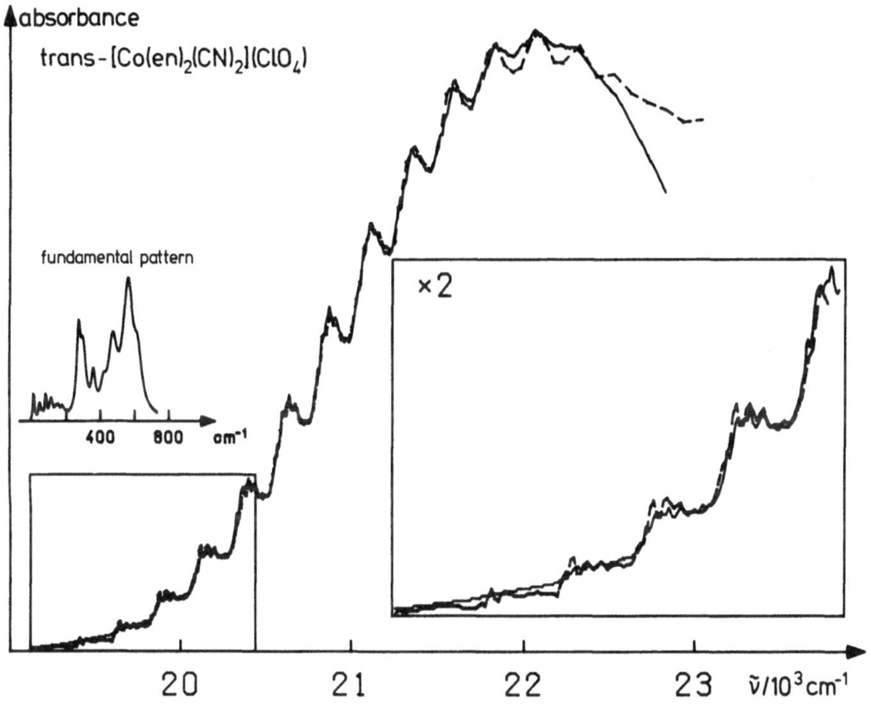

Fig. 16. Experimental spectrum of *trans*-[$Co(en)_2(CN)_2$]ClO_4 polarized closely parallel to the crystal b axis (*dashed line*) and calculated from superpositions of the fundamental spectral pattern as depicted in the *inset* correcting for the band overlap due to the higher singlet transitions (*solid line*)

been carried out for a size reduced molecule in which the NH_2 and CH_2 groups of ethylenediamine are substituted by N* and C* pseudo atoms of corresponding masses [97]. The resulting normal coordinate belonging to the vibrational mode which builds up the progression is

$$Q(a_g) = 0.44 S_1 - 0.47 S_2 - 0.56 S_9 + 0.47 \Delta S_{10}$$

when S_1 and S_2 are Co–N and N–C stretching and S_9 and S_{10} NCoN angular symmetry vibrations, respectively. With the reduced mass $M_{a_{1g}} = 37.04$ amu, also obtained from the normal coordinate analysis, for use in Eq. (18), the molecular constants are

$$\Delta Q(a_{1g}) = 29.8 \text{ pm},$$

$$A_{a_g}^{1B_g} = -3.7. \times 10^{-9} \text{ J/m}$$

$$\Delta E_{a_g} = -2700 \text{ cm}^{-1}$$

With this ΔQ value the change in the symmetry coordinate ΔS_1 of pure Co–N stretching is calculated leading to alternations of bond distances in the x, y plane of

$$\Delta x_1 = \Delta x_2 = \Delta y_3 = \Delta y_4 = 6.6 \text{ pm}$$

In addition, also the equilibrium NCoN angle is changed due to the contribution of these symmetry coordinates in the $Q(a_g)$ coordinate given above. The deuterated complex treated in similar way has almost equal molecular constants although, as we have seen in the corresponding complex with NH_3 ligands, their vibrational constants are quite different [97].

As a last example we want to look at Pt(IV) halogen complexes of the type A_2PtX_6 where X = F, Cl and Br and A = K or Cs. The emission from the lowest excited state $\Gamma_3(^3T_{1g})$ of the chloride complex shows a progression in $\nu_2(e_g)$ with a substructure in which a combination with the three promoting modes ν_3, ν_4 and ν_6 of odd vibrations of the $PtCl_6^{2-}$ octahedron becomes apparent which make the *d–d* transition parity allowed (cf. Fig. 17). The progression in steps of 310–323 cm^{-1} compares well with $\nu_2(e_g) = 320\ cm^{-1}$ measured in the Raman spectrum [98] indicating a Jahn-Teller effect in the Γ_3 excited state by $E \times e$ coupling. The spectroscopic constants obtained from the least-square fit and the molecular parameters calculated thereof are listed in the attached table (taken from Ref [19]).

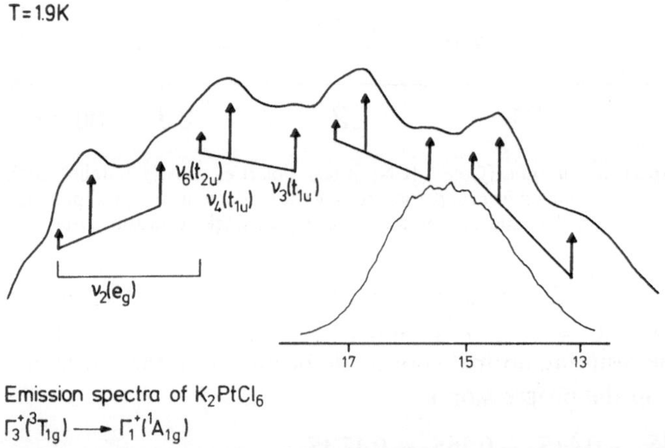

Fig. 17. Emission spectrum of a K_2PtCl_6 single crystal at T = 1.9 K (section at absorption maximum). Excitation wavelength $\lambda_{exc} = 454.5$ nm from an argon ion laser. The *inset* is the overall spectrum, the energy plotted in units of $kK = 10^3\ cm^{-1}$

Table 1. Spectroscopic and molecular constants of octahedral A_2PtX_6 complexes ($E \times e$ coupling case) for $\Gamma_3(^3T_{1g}) \rightarrow \Gamma_1(^1A_{1g})$ emission

	Cs_2PtF_6	K_2PtCl_6	K_2PtBr_6
$\hbar\omega$ (cm^{-1})	566	310	190
β	1	1	0.94
Δ	3.5	5.4	5.1
ΔQ (pm)	20	30	24
A ($Jm^{-1} \times 10^{-9}$)	14	12	8
ΔE (cm^{-1})	3500	4500	2500
E_{00} (cm^{-1})	18350	19500	16100

Obviously the parameters do not follow unique trends in the series of halogen complexes indicating that vibronic interaction in the molecules is determined by several contributing factors. In this context it should be referred to the calculation of molecular coupling parameters on the basis of dynamic ligand field theory using common ligand field parameters [19]. In this model the ligands are not fixed but allowed to vibrate relative to the central metal, and the *d*-electron energy levels are subjected to perturbation by the linear vibronic coupling operator. We shall, however, not go into further detail. Instead, we would like to draw attention to the complications due to the energetic neighborhood of other electronic states of the emitting Γ_3 level. As a matter of fact, there are several close to it. Group theory predicts, e.g., a splitting of the $^3T_{1g}$ level (lowest of $t_{2g}^5 e_g$ configuration) into four spin-orbit levels, i.e., $\Gamma_1, \Gamma_3, \Gamma_4$ and Γ_5 (Bethe notation [99]). Ligand field calculations performed with reasonable model parameter sets result in three closely related levels, which are Γ_3 (lowest), Γ_5 and Γ_4 differing by only about 1000 cm^{-1} in energy, much independent of the spin-orbit coupling parameter chosen [100]. It is to be expected that these levels will strongly intermix by vibronic coupling, e.g., by interaction with t_{2g} vibrational modes generating, as in Eq. (4), a nondiagonal element

$$\left\langle \Psi(\Gamma_3) \middle| \left(\frac{\partial \mathrm{H}}{\partial \mathrm{Q}_{t_{2g}}} \right)_0 \middle| \Psi(\Gamma_5) \right\rangle \neq 0 \tag{61}$$

This intermixing will distort the potential hypersurface in the t_{2g} subspace of the octahedral vibrational coordinates. In a similar way, maybe less effective, the lowest Γ_3 hypersurface will be scrambled in the e_g subspace by interaction with Γ_1, Γ_2 and Γ_3 levels, the latter two resulting from $^3T_{2g}$ due to splitting by spin-orbit coupling. We are well familiar with this situation which is the same as in the case described for s^2 systems [38] (cf. Sect. 3.2): the perturbation matrix of the vibronic operator in e_g and t_{2g} space for the two lowest $t_{2g}^5 e_g$ triplets is 18 dimensional, its Q-dependent eigenvalues distort the potential hypersurfaces of the 8 possible (by group theory) spin-orbit levels due to the 1st and 2nd order Jahn-Teller effect. For the lowest $\Gamma_3(^3T_{1g})$ level we expect a potential in the e_g space which does not have any more the simple shape of a "Mexican hat" [54] but may be distorted eventually forming simultaneous side minima which can have a remarkable influence on the electronic spectra.

This, for instance, may explain the unexpected observation that the low temperature (2 K) absorption and emission spectra of K_2PtCl_6 overlap within an energy region as large as about 4500 cm^{-1} [30, 32, 82]. While from the well resolved emission spectrum (Fig. 17) the zero-phonon line of the $\Gamma_3(^3T_{1g}) \rightarrow \Gamma_1(^1A_{1g})$ transition is extrapolated in a good approximation to be at 19 500 cm^{-1} [82], the absorption spectrum has a weak but distinct band feature starting at about 15 000 cm^{-1} with a maximum at 18 000 cm^{-1} [32]. If this absorption band is due to the same molecular species and is assigned to the same transition, there must be two different origins attributed to the excited state potential. This can be realized if the lower Γ_3 hypersurface has two minima in the e_g space, a main (absolute) minimum and a higher side (relative) minimum

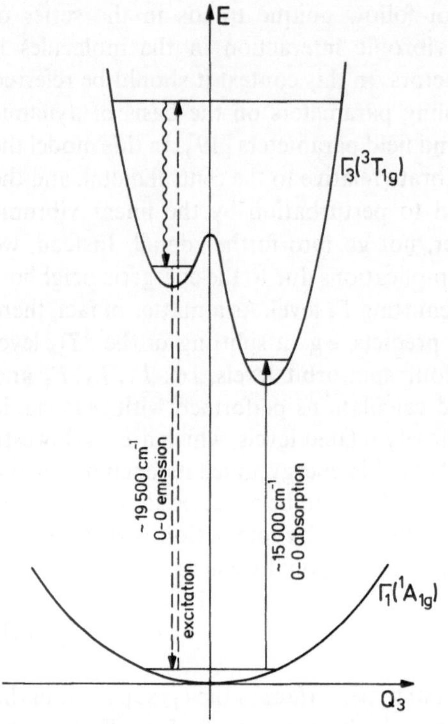

Fig. 18. Possible potential hypersurface of the $\Gamma_3(^3T_{1g})$ state of $PtCl_6^{2-}$ in the $e_g(Q_2, Q_3)$ space as projected into the Q_3 energy plane (the side minimum may represent only a saddle point)

(cf. Fig. 18). In absorption the transition goes from the ground state into the absolute minimum of the excited state potential, in emission the excitation energy is collected in the side minimum from where it emits light before it can relax into the lower absolute minimum. The particular shape of the Γ_3 hypersurface depicted in Fig. 18, which can explain the observed optical behavior, is formed by vibronic coupling of higher electronic levels into the Γ_3 level. This is, of course, only one possible explanation of experimental findings which does not claim uniqueness.

4 Conclusions

An explanation of band profiles of electronic absorption or emission spectra from transition probabilities must also include vibrational transitions. In general, electron excitation (or deactivation) is assisted by changes in vibrational excitation. The simultaneous alternation of electron and nuclear motion occurring during a transition between energy levels is described by vibronic coupling, in general using the Herzberg-Teller coupling scheme. On the basis of this theory one can carry out a Franck-Condon analysis of experimental band profiles by fitting measured intensities to theoretical band shape functions.

The present article compiles, for the first time, an extensive theoretical approach to this band analysis reported at various places in the literature which is the most comprehensive procedure available at present time. It includes cases where potential curves of excited states are shifted in configuration space relative to the ground state, when force constants and fundamental frequencies belonging to these states are different (frequency effect), if the electronic transition is symmetry forbidden such that promoting (active) modes making the transition possible must be involved. Moreover, the theory is able to consider degenerate modes which give rise to a progression, and it can describe the temperature dependence of the intensity distribution in the progression by occupation of higher vibrational levels of the excited state. The model allows us to consider the anti-Stokes part of spectrum, as well, and it may be extended including the Duschinsky effect [67]. Since all these cases can be considered simultaneously, the present method is to a great extent unique and superior to earlier Frank-Condon procedures.

In the experimental section the use of the theory is demonstrated by reporting some results on several systems for which it has been applied. Since the final theoretical formulas are in general given in an analytical form, a least square fit to the band profiles of the spectra is not difficult and can be carried out in a straightforward way. The model parameters obtained are coupling constants, vibrational frequencies and atomic distances for the excited state, zero-phonon transitions and Jahn-Teller stabilization energies etc. These results supply important data for gaining valuable insight on the chemical bond, the reactivity and kinetics of the ground and excited states. The error limits in the parameter values are primarily determined by the quality of the spectra obtained by experiment. An improvement of experimental techniques aiming to get better resolved spectra will have a large impact on the value of the present method. Therefore, an evaluation of the results depends in the first place on the spectra available.

Acknowledgment. The author is grateful to Privatdozent Dr. Joachim Degen, Düsseldorf, for many interesting discussions concerning vibronic coupling. All figures copied from the literature are with the permission of the respective publishers.

5 References

1. Jacobsen SM, Reber C, Güdel HU (1988) J Luminesc 40, 41: 107
2. Reber C, Güdel HU, Meyer G, Schleid T, Daul CA (1989) Inorg Chem 28: 3249
3. Gälli B, Hauser A, Güdel HU (1985) Inorg Chem 24: 2271
4. Winter NW, Pitzer RM (1988) J Chem Phys 89: 446
5. Lever ABP (1984) Inorganic electronic spectroscopy, 2nd edn. Elsevier, Amsterdam
6. Hoggard PE (1986) Coord Chem Rev 70: 85

7. Flint CD (1974) Coord Chem Rev 14: 47
8. Luk CK, Yeakel WC, Richardson FS, Schatz PN (1975) Chem Phys Lett 34: 147
9. Patterson HH, Hasan Z, Manson NB (1980) Chem Phys Lett 75: 156
10. Flint CD, Paulusz AG (1981) Mol Phys 44: 925
11. Flint CD, Lang P (1981) J Luminesc 24, 25: 301
12. Black AM, Flint CD (1977) J Chem Soc Faraday Trans II 73: 877
13. Wernicke R, Schmidtke H-H (1979) Mol Phys 37: 607
14. Bettinelli M, Di Sipio L, Ingletto G, Montenero A, Flint CD (1985) Mol Phys 56: 1033
15. Bellitto C, Brunner H, Güdel HU (1987) Inorg Chem 26: 2750
16. Khan SM, Patterson HH, Engstrom H (1978) Mol Phys 35: 1623
17. Kozikowski BA, Keiderling TA (1980) Mol Phys 40: 477
18. Strand D, Linder R, Schmidtke H-H (1990) Mol Phys 71: 1075
19. Schmidtke H-H, Degen J (1989) Struct and Bdg 71: 99
20. Valek MH, Yeranos WA, Basu G, Hon PK, Belford RL (1971) J Mol Spectry 37: 228
21. Garner CD, Hill LH, Mabbs FE, McFadden DL, McPhail AT (1977) J Chem Soc Dalton 853
22. Sartori C, Preetz W (1988) Z Naturforsch 43a: 239
23. Güdel HU, Snellgrove TR (1978) Inorg Chem 17: 1617
24. Chen MY, McClure DS, Solomon DS (1972) Phys Rev B 6: 1690, 1697
25. Douglas IN (1969) J Chem Phys 51: 3066
26. Schmidtke H-H, Göttges D (1989) Ber Bunsenges Phys Chem 93: 135
27. Wilson RB, Solomon EI (1980) J Am Chem Soc 102: 4085
28. Komi Y, Urushiyama A (1980) Bull Chem Soc Jpn 53: 980
29. Hipps KW, Merrell GA, Crosby GA (1976) J Phys Chem 80: 2232
30. Eyring G, Schmidtke H-H (1981) Ber Bunsenges Phys Chem 85: 603
31. Patterson HH, De Berry WJ, Byrne JE, Hsu MT, LoMenzo JA (1977) Inorg Chem 16: 1698
32 Yoo RK, Keiderling TA (1990) J Phys Chem 94: 8048
33. Hamm DJ, Schreiner AF (1975) Inorg Chem 14: 519
34. Vehse WE, Lee KH, Yun SI, Sibley WA (1975) J Luminesc 10: 149
35. Hitchman MA, Cassidy PJ (1979) Inorg Chem 18: 1745
36. Rice SF, Gray HB (1983) J Am Chem Soc 105: 4571
37. Bär L, Englmeier H, Gliemann G, Klement U, Range K-J (1990) Inorg Chem 29: 1162
38. Wernicke R, Kupka H, Enßlin W, Schmidtke H-H (1980) Chem Phys 47: 235
39. Donker H, Smit WMA, Blasse G (1989) J Phys Chem Solids 50: 603
40. Harrison TG, Patterson HH, Godfrey JJ (1976) Inorg Chem 15: 1291
41. Patterson HH, Godfrey JJ, Khan SM (1972) Inorg Chem 11: 2872
42. Di Sipio L, Oleari L, Day P (1972) J Chem Soc Faraday Trans II 68: 776
43. Collingwood JC, Schwartz RW, Schatz PN, Patterson HH (1974) Mol Phys 27: 1291
44. Collingwood JC, Piepho SB, Schwartz RW, Dobosh PA, Dickinson JR, Schatz PN (1975) Mol Phys 29: 793
45. Collingwood JC, Schatz PN, McCarthy PJ (1975) Mol Phys 30: 469
46. Piepho SB, Dickinson JR, Spencer JA, Schatz PN (1972) Mol Phys 24: 609
47. Piepho SB, Dickinson JR, Spencer JA, Schatz PN (1972) J Chem Phys 57: 982
48. Kroening RK, Rush RM, Martin DS, Clardy JC (1974) Inorg Chem 13: 1366
49. Schwarz RW (1977) Inorg Chem 16: 836
50. Johnson LW, McGlynn SP (1970) Chem Phys Lett 7: 618
51. Holt SL, Ballhausen CJ (1967) Theoret Chim Acta 7: 313
52. Clark RJH, Dines TJ, Wolf ML (1982) J Chem Soc Faraday Trans II 78: 679
53. Braun D, Gallhuber E, Hensler G, Yersin H (1989) Mol Phys 67: 417
54. Braun D, Hensler G, Gallhuber E, Yersin H (1991) J Phys Chem 95: 1067
55. Bersuker IB, Polinger VZ (1989) Vibronic interactions in molecules and crystals. Springer, Berlin Heidelberg New York
56. Ballhausen CJ, Hansen AE (1972) Annu Rev Phys Chem 23: 15
57. Ballhausen CJ (1979) Molecular electronic structures of transition metal complexes. McGraw-Hill, New York
58. Kupka J (1976) Chem Phys Lett 41: 114
59. Kupka H, Enßlin W, Wernicke R, Schmidtke H-H (1979) Mol Phys 37: 1693
60. Lin SH (1966) J Chem Phys 44: 3759
61. Kupka H (1978) Mol Phys 36: 685
62. Kupka H (1979) Mol Phys 37: 1673
63. Kupka H, Schmidtke H-H (1981) Mol Phys 43: 451

64. Nitzan A, Jortner J (1973) Theor Chim Acta 30: 217
65. Degen J, Kupka H, Schmidtke H-H (1987) Chem Phys 117: 163
66. Kupka H (1980) Mol Phys 39: 849
67. Duschinsky F (1937) Acta Phys Chim USSR 7: 551
68. Olbrich G, Kupka H (1983) Z Naturforsch 38a: 937
69. Kupka H, Cribb PH (1986) J Chem Phys 85: 1303
70. Wilson RB, Solomon EI (1978) Inorg Chem 17: 1729
71. Flint CD, Matthews AP (1975) Inorg Chem 14: 1219
72. Ölkrug D, Radjaipour M, Eitel E (1979) Spectrochim Acta 35A: 167
73. Heller EJ (1978) J Chem Phys 68: 3891
74. Tannor DJ, Heller EJ (1982) J Chem Phys 77: 202
75. Tutt L, Tannor D, Schindler J, Heller EJ, Zink JI (1983) J Phys Chem 87: 3017
76. Ranfagni A, Mugnai D, Bacci M, Viliani G, Fontana MP (1983) Adv Phys 32: 823
77. Oomen EWJL, Smit WMA, Blasse G (1984) Chem Phys Lett 112: 547
78. Donker H, Van Schaik W, Smit WMA, Blasse G (1989) Chem Phys Lett 158: 509
79. Schmidtke H-H, Krause B, Schönherr T (1990) Ber Bunsenges Phys Chem 94: 700
80. Stufkens DJ (1970) Rec Trav Chim 89: 1185
81. Meidenbauer K, Gliemann G (1988) Z Naturforsch 43a: 555
82. Schmidtke H-H (1986) In: Lever ABP (ed) Excited states and reactive intermediates, ACS Symposium Series 307: 23
83. Blasse G, Dirksen GJ, Abriel W (1987) Chem Phys Lett 136: 460
84. Fukuda A (1970) Phys Rev B1: 4161
85. Degen J, Schmidtke H-H (1989) Chem Phys 129: 483
86. Hesse K, Gliemann G (1991) J Phys Chem 95: 95
87. Schmidtke H-H, Diehl M, Degen J (1992) J Phys Chem 96: 3605
88. Degen J, Diehl M, Schmidtke H-H (1993) Mol Phys 78: 103
89. Jacobs PWM, Oyama K (1975) J Phys C 8: 851, 865
90. Kamishina Y, Sivasankar VS, Jacobs PWM (1982) J Chem Phys 76: 4677
91. Grzonka C (1992) Doctoral thesis university Düsseldorf
92. Schmidtke H-H, Schönherr T (1990) Chem Phys Lett 168: 101
93. Yeakel WC, Schatz PN (1974) J Chem Phys 61: 441
94. Yeakel WC, Slater JL, Schatz PN (1974) J Chem Phys 61: 4868
95. Preetz W, Tensfeldt D (1984) Z Naturforsch 39a: 966
96. Urushiyama A, Kupka H, Degen J, Schmidtke H-H (1982) Chem Phys 67: 65
97. Hakamata K, Urushiyama A, Degen J, Kupka H, Schmidtke H-H (1983) Inorg Chem 22: 3519
98. Woodward LA, Creighton JA (1961) Spectrochim Acta 17: 594
99. Bethe H (1929) Ann Phys [5] 3: 133
100. Eyring G, Schönherr T, Schmidtke H-H (1983) Theor Chim Acta 64: 83

Sharp-Line Electronic Spectra and Metal-Ligand Geometry

Patrick E. Hoggard

Department of Chemistry, North Dakota State University, Fargo, ND 58105, USA

Table of Contents

Topics in Current Chemistry, Vol. 171

Patrick E. Hoggard

The narrow intraconfigurational lines in the electronic spectra of transition metal complexes can, unlike broad bands, be assigned to single electronic transitions. The positions of, and especially the splittings within, these lines are strongly influenced by the angular positions and orientations of the ligands. The angular overlap model (AOM) is well suited to treat the exact geometry as part of the ligand field potential. It is possible in some instances to deduce particular bond angles by analysis of the electronic spectrum. It is suggested that when AOM parameters are deduced from spectral data, the ligand geometry be explicitly included, either directly from an X-ray crystal structure, if available, or from an approximate method, such as a molecular mechanics calculation. This ought to insure the maximum transferability of parameters for a particular ligand in different complexes, and for a particular coordinating group in different ligands.

1 Introduction

Ligand field theory has proved to be very powerful in the extraction of some kinds of structural information from the electronic spectra of transition metal complexes. Oxidation state, total coordination number, the coordination number for each different ligating atom, and the geometric isomer at hand are among these. It has been much more difficult to derive geometric details such as bond angles, even though the Angular Overlap Model (AOM) formulation of the ligand field potential [1, 2] clearly depends directly on the orientation of the ligating atoms around the metal.

There are three main reasons why bond angle information generally remains buried in electronic spectra: the bands observed are too few, too broad, and too complex. Band positions are affected by ligand field strengths, interelectronic repulsion, and spin-orbit coupling, as well as bond angles, and there must be enough bands to uniquely determine all the factors that go into the model. Large band widths mean sizable uncertainties in the position, which are usually propagated into still larger uncertainties in the parameters derived in a fitting process. Most broad bands also consist of several unresolved components. The $^1A_{1g} \rightarrow {}^1T_{1g}$ transition in low-spin Co(III) complexes, for example, comprises three components, while the $^4A_{2g} \rightarrow {}^4T_{2g}$ transition in Cr(III) complexes comprises six. The association of a band position with a calculated transition energy, or an average calculated transition energy, becomes very problematic.

All three of these problems tend to be resolved when bands due to intraconfigurational transitions are used. In six-coordinate, approximately O_h complexes these are $t_{2g} \rightarrow t_{2g}$ and $e_g \rightarrow e_g$ transitions, quite often accompanied by a spin flip, and therefore forbidden. In contrast to $t_{2g} \rightarrow e_g$ transitions, these are usually numerous, narrow (because the excited state resembles the ground state), and simple: each observed narrow line comprises just one component. The relative abundance of these sharp lines (8 for d^3 complexes, for example, and as many as 14 for low-spin d^4) is deceptive: it is because the also numerous $t_{2g} \rightarrow e_g$ components cluster under broad bands.

There are experimental disadvantages that detract from the advantages just listed. The extinction coefficients may be so low that lines are obscured even by remote tails from spin-allowed transitions. The transition energies may span a wide range, requiring UV-visible, near-IR, and mid-IR instrumentation. Electronic lines may also be accompanied by many vibrational satellites whose intensities are high enough to make the distinction between electronic and vibronic lines very difficult.

In our work we have concentrated heavily on chromium(III) complexes. The eight sharp lines are nearly always in the visible spectral range, and divide into a group of five peaks ($^4A_{2g} \rightarrow {}^2E_g, {}^2T_{1g}$ in O_h notation) in the red, frequently well removed from the nearest spin-allowed band, and a group of three ($^4A_{2g} \rightarrow {}^2T_{2g}$) that often falls between the first and second spin-allowed bands, but may be obscured by them. Most Cr(III) complexes luminesce at low temperature, which makes it possible to record these peaks in excitation spectra, enhancing the sensitivity over absorption spectra.

The central features of the Angular Overlap Model are (1) the use of the sigma- and pi-destabilization parameters, e_σ and e_π, to express the geometry-independent effects of a ligand on the metal d orbitals; (2) a mechanism to evaluate those effects for a ligand at any angular position with respect to a defined set of d orbitals [1,2]; and (3) the ability to sum the effects from all ligands, or indeed any other perturbers whose angular positions can be specified. The parameter e_σ is defined as the increase in energy of the d_{z^2} orbital through coordination of a ligand on the z-axis. A consequence of this definition is that a ligand on the x- or y-axis will raise the energy of the $d_{x^2-y^2}$ orbital by $3e_\sigma/4$ and the energy of the d_{z^2} orbital by $e_\sigma/4$. The parameter e_π is defined as the amount by which the d_{xz} or d_{yz} orbital increases in energy through coordination of a ligand on the z-axis. The interaction between a ligand σ lone pair and a metal d orbital should always stabilize the lone pair orbital and destabilize the metal d orbital, so e_σ should always be positive. The π-interaction should destabilize a metal d orbital if it is with a ligand lone pair orbital, but should stabilize the d orbital if the interaction is with an empty π^* orbital. Both can occur at the same time, so e_π will be positive or negative, depending on whether the predominant interaction is with a bonding or an antibonding orbital, respectively.

The AOM parameters are often interpreted in terms of the donor and acceptor properties of ligands. Large positive values of e_σ and e_π characterize strong donors, while negative values of e_π characterize π-acceptors. To some degree, e_σ and e_π values should be constant in related complexes with the same metal ion, a point that will be addressed in Sect. 8. As might be expected, e_σ values tend to be considerably larger than e_π values. Properties that depend on the total donor ability of a ligand might be expressed in terms of $e_\sigma + e_\pi$. The ratio of e_σ to e_π is expected to stay relatively constant from one metal ion to another, except when a filled subshell is encountered. Six-coordinate low spin d^6 complexes, for example, should have only a limited ability to accept electrons into the t_{2g} orbitals, which should reduce considerably the donor strength of otherwise strongly π-donating ligands.

The means by which the AOM is used to construct the 5×5 ligand field potential matrix, $\langle d_i|V|d_j\rangle$, in terms of the e_σ and e_π parameters, have been presented in some detail [3]. This 1-electron matrix is then used to generate the 3-electron secular equation from a Hamiltonian that includes interelectronic repulsion, through the Racah parameters B and C, and the Trees correction parameter α_T, and spin-orbit coupling [3]. Schmidtke has also shown that the spherical interelectronic repulsion parameters are inadequate to describe complexes with approximate tetragonal symmetry when there is a π-interaction [4]. In such cases we use the Schmidtke τ factor as a fixed correction [5, 6]. The eigenvalues of the secular matrix can be fit to observed transition energies, and the best values of the parameters from the model, including any angular variables used to construct the $\langle d_i|V|d_j\rangle$ matrix, can be found.

2 Angular Variables and the Ligand Field Potential Matrix

All of the metal–ligand bonding characteristics (to the extent of the ligand field approximation), the symmetry of the molecule, both actual and approximate, and the precise disposition of the ligands around the metal ion are tied up in the 5×5 ligand field potential matrix. Contributions to these matrix elements arise not just from the coordinated atoms, but also from atoms further away, whose symmetry may not be the same as that of the ligating atoms [7]. Traditional ligand field theory is defined around this matrix. The parameters to be used in spectral fitting are usually chosen as the simplest set necessary to define the matrix, given the actual or an approximate molecular symmetry.

For example, in O_h symmetry, the familiar matrix has the form

$$\begin{matrix} -4Dq & 0 & 0 & 0 & 0 \\ 0 & -4Dq & 0 & 0 & 0 \\ 0 & 0 & -4Dq & 0 & 0 \\ 0 & 0 & 0 & 6Dq & 0 \\ 0 & 0 & 0 & 0 & 6Dq \end{matrix} \qquad \text{(M1)}$$

The sequence of rows and columns in the $\langle d_i|V|d_j\rangle$ matrices in this paper is d_{xy}, d_{xz}, d_{yz}, $d_{x^2-y^2}$, d_{z^2}. One parameter is necessary to determine all energy differences. Choosing the diagonal elements so that the center of gravity is zero is, of course, unnecessary for the calculation of transition energies.

In the AOM formalism the ligand field potential matrix in O_h symmetry is

$$\begin{matrix} 4e_\pi & 0 & 0 & 0 & 0 \\ 0 & 4e_\pi & 0 & 0 & 0 \\ 0 & 0 & 4e_\pi & 0 & 0 \\ 0 & 0 & 0 & 3e_\sigma & 0 \\ 0 & 0 & 0 & 0 & 3e_\sigma \end{matrix} \qquad \text{(M2)}$$

This matrix is arrived at by summing the contributions from the six ligands. For example, the last term, $\langle d_{z^2}|V|d_{z^2}\rangle$, arises from an increase in energy of e_σ caused by each of the two ligands on the z-axis, plus $e_\sigma/4$ from the four ligands in the xy-plane, to give a total destabilization of $3e_\sigma$. Two parameters are necessary to specify the matrix, but since there is only one energy difference ($3e_\sigma - 4e_\pi$), e_σ and e_π cannot be separately determined from spectroscopic measurements.

Where there are two different ligating atoms, there are four AOM parameters, or fewer if one or both of the ligands is a saturated amine, so that π-interactions can be neglected. There may also be angular variables necessary to express the atom positions. Again, a comparison with the traditional ligand field theory approach is useful. For a *trans*-MA_4B_2 complex, with coordinated atoms on the Cartesian axes, the ligand field potential matrix takes the general form

$$\begin{matrix} a & 0 & 0 & 0 & 0 \\ & b & 0 & 0 & 0 \\ & & b & 0 & 0 \\ & & & c & 0 \\ & & & & d \end{matrix} \qquad \text{(M3)}$$

In the AOM formalism, a is $4e_{\pi A}$, b is $2e_{\pi A} + 2e_{\pi B}$, c is $3e_{\sigma A}$, and d is $e_{\sigma A} + 2e_{\sigma B}$. Again, since only energy differences are determined spectroscopically, there are only three recoverable degrees of freedom. We denote this by saying that matrix M3 has three spectroscopically independent parameters. In crystal field theory these are defined as

$$Dq = \frac{1}{10}(c - a) \qquad (1)$$

$$Ds = \frac{1}{7}(a - b + c - d) \qquad (2)$$

$$Dt = \frac{1}{35}(-4a + 4b + 3c - 3d) \qquad (3)$$

In these examples the classical and AOM approaches are equivalent, but the AOM parameters cannot be determined uniquely unless A or B is a saturated amine (ignoring the question of obtaining the requisite spectroscopic data). Otherwise there are four independent AOM parameters, which is one too many.

Approximate tetragonal symmetry is often also assumed for chelate complexes of the type *trans*-$M(A{-}A)_2B_2$. Without this assumption, but specifying that the B ligands have a cylindrically symmetric π-interaction with the metal, the ligand field potential matrix has the general form

$$\begin{array}{ccccc} a & 0 & 0 & 0 & f \\ & b & e & 0 & 0 \\ & & b & 0 & 0 \\ & & & c & 0 \\ & & & & d \end{array} \qquad \text{(M4)}$$

The assumption of tetragonal symmetry means ignoring the e and f off-diagonal terms, whose magnitudes are determined by up to three angular variables. These may be defined [3] as the Cartesian bite angle (α), the angular displacement of the A atoms from the Cartesian axes; the Cartesian twist angle (β), the angle by which the plane defined by an M–A–A triangle is rotated out of the xy, xz, or yz plane; and the π-orientation angle (ψ), which describes the position of the orbital used for π-bonding on an A atom [3]. The Cartesian bite and twist angles are illustrated in Fig. 1. If the M–B bonding were not cylindrically symmetric, i.e., if the π-bonding were anisotropic, the degeneracy of the b terms would be lifted, and there would be seven degrees of freedom in the matrix.

The *trans*-$M(A{-}A)_2B_2$ matrix M4 has six degrees of freedom and five spectroscopically independent elements. It is certainly within the realm of the

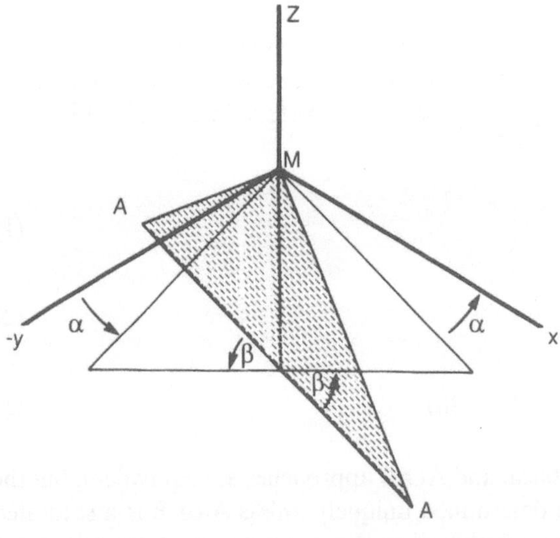

Fig. 1. The Cartesian bite angle α is the angular displacement of the metal-ligand bond from an axis. The Cartesian twist angle β is the angular displacement of the M–A–A triangle from its reference plane. If the Cartesian axes are not uniquely defined, they can be improvised so as to minimize the squares of the displacements from the axes

possible to determine these from spectral data, if enough sharp-line electronic transitions can be located and correctly identified. Determining the four AOM parameters ($e_{\sigma A}$, $e_{\sigma B}$, $e_{\pi A}$, $e_{\pi B}$) and three angular variables from the five energy differences (or directly from a fit to the spectroscopic data, without first finding the $\langle d_i|V|d_j\rangle$ matrix elements) is another matter. Clearly it is impossible to determine seven parameters uniquely, given only five spectroscopically independent elements.

The parameter space may be reduced, however, in some circumstances. If A is a saturated amine, then $e_{\pi A}$ is zero, and the orientation angle ψ drops out of the problem. The ligand field potential matrix still has the form of matrix M4, with six terms, but now the five spectroscopically independent elements are sufficient to determine uniquely the five remaining parameters ($e_{\sigma A}$, $e_{\sigma B}$, $e_{\pi B}$, α, β). The e term in matrix M4 is directly related to the Cartesian twist angle. If that angle is zero, e is zero, and the four remaining parameters, including the bite angle α, can be determined from the five nonzero matrix elements (four spectroscopically independent).

The key question is what significance can be attached to parameters, specifically AOM parameters and angular variables, derived from spectroscopic data. As these examples were intended to illustrate, it is useful to divide this question into two parts, whether or not a calculation is actually performed this way: (1) can the spectroscopically independent elements of the ligand field potential matrix be uniquely determined from the data? (2) can the AOM parameters and angular variables be uniquely determined from the spectroscopically independent elements of the potential matrix?

Qualitatively, the first question can be answered affirmatively if enough electronic peaks are found and correctly assigned. Along with the sharp-line peaks, the broad bands may be included, with perhaps a deconvolution to resolve underlying structure. Even with deconvolution, broad band maxima must usually be assigned to an average of several different transitions, causing additional uncertainties. Uncertainties in experimental peak positions, along with imputed uncertainties in assignments of one peak to several transitions, can be used as weighting factors in the evaluation of the errors propagated to the parameter values that are determined in the fitting process [8]. The requirement is that the number of assigned experimental peak positions equal or (preferably) exceed the number of spectroscopically independent elements in the ligand field potential matrix, plus the number of additional parameters for interelectronic repulsion and spin–orbit coupling. This is typically four in our work (B, C, α_T, and ζ; see Sect. 1).

The obvious answer to the second part is that the number of spectroscopically independent elements in the ligand field potential matrix is the upper limit to the number of AOM parameters and angular variables that may be determined from them. No matter how many experimental peak positions have been measured, it is the form of the ligand field potential matrix that determines how much information can be extracted. If there are more parameters than spectroscopically independent matrix elements, it may still be possible to determine some parameters directly or in useful linear combinations, such as $3e_\sigma - 4e_\pi = 10Dq$. Diagonal elements are always relative; a common constant can be added without affecting transition energies. Off-diagonal elements, however, are absolute in magnitude but arbitrary in sign, and any parameters that can be deduced directly from them are valid (except possibly for sign) regardless of the dimensionality of the rest of the model.

When the number of AOM parameters plus angular variables is greater than the number of spectroscopically independent elements in the potential matrix it may still be possible to determine all the parameters by placing restrictions on their values, provided the restrictions can be justified. An examination of the propagated uncertainties will reveal whether this device is successful or not. Our experience has been that the uncertainties in the angular values are unacceptably high when there are too few spectroscopically independent elements in the potential matrix.

Thus, the 5×5 ligand field potential matrix is the key to the acquisition of meaningful information. For some high symmetry situations traditional ligand field theory has defined parameters that are linear combinations of these matrix elements [9]. Some of them have chemical significance, while others do not. Transferability to other complexes, particularly complexes with lower symmetry or different angular geometry, is quite problematic.

The AOM formalism was designed to address the problem of transferability, and the parameters have specific and considerable chemical significance. However, as seen from the example above, problems must be carefully chosen if any substance is to be attached to the values derived. Sometimes (but not always), because the baseline for the diagonal elements is arbitrary, one of those elements (or sometimes a parameter that contributes to two or more elements) must be known with some confidence. This is why saturated amine complexes are often chosen for study, allowing $e_{\pi N}$ to be set to zero.

A question that sometimes arises in this regard is whether the dimensionality of the ligand field potential matrix, as dictated by symmetry, is actually smaller than what is apparent by inspection of the matrix itself. In other words, are there hidden redundancies in the matrix? It is sometimes thought that four is the maximum number of spectroscopically independent parameters that can be extracted from spectral data, because there are just four energy differences among the d orbitals. This would be true if it were not for interelectronic repulsion. The many-electron ligand field energies can be calculated from the one-electron eigenvalues of the ligand field potential matrix. This is the strong field approximation in the limit. But the splittings of, and energy contributions to, these states due to interelectronic repulsion depend on the orbital composition of each state, that is, on the eigenvectors of the ligand field potential matrix. The eigenvectors of a general 5×5 symmetric matrix have ten degrees of freedom (this is because the 5×5 orthonormal matrix of the eigenvectors (O) can be put in correspondence with a skew-symmetric matrix (S) as $O = e^S$ [11]; there are 10 distinct off-diagonal elements in a 5×5 skew-symmetric matrix). Together with the five degrees of freedom from the eigenvalues this is equivalent to the fifteen degrees of freedom in the original symmetric ligand field potential matrix. Less the arbitrary constant on the diagonal, there are, in principle, fourteen degrees of freedom that could be determined from spectral data, however unlikely this might be in practice. Note that some sets of off-diagonal elements may actually define rotations in three-space, to which the energies are invariant. If such a case were to arise it would be impossible to determine the off-diagonal elements spectroscopically. However, this does not in any way reduce the dimensionality (15) of the general case.

To illustrate this point, take an $MA_2B_2C_2$ complex in the all-*trans* geometry, with D_{2h} symmetry. The ligand field potential matrix has the general form

$$\begin{array}{ccccc} a & 0 & 0 & 0 & 0 \\ & b & 0 & 0 & 0 \\ & & c & 0 & 0 \\ & & & d & f \\ & & & & e \end{array} \qquad \text{(M5)}$$

The six elements can be used to generate the n-electron ligand field energies, to which interelectronic repulsion and spin-orbit coupling, which are already defined on the $\{d_{xy}, d_{xz}, d_{yz}, d_{x^2-y^2}, d_{z^2}\}$ basis, are added. Alternatively, the five eigenvalues can be used, but an additional term is necessary to define the correct linear combinations of the $d_{x^2-y^2}$ and d_{z^2} orbitals on which to evaluate interelectronic repulsion and spin-orbit coupling. Either way, there are still six total, or five spectroscopically independent, terms in the ligand field potential matrix.

Symmetry-adapted d orbitals can sometimes be specified in advance, so that the dimensionality of the matrix can be precisely defined by the degeneracy of the d orbitals. Then, and in intermediate symmetries, there may indeed be redundancy in the potential matrix. This might be uncovered by inspection. For example, when $e_{\sigma A} = e_{\sigma B}$ and $e_{\pi A} = e_{\pi B} = 0$, and $\alpha = 0$ but β is nonzero, matrix M4 above for *trans*-$M(A\text{–}A)_2B_2$ assumes the form

$$\begin{array}{ccccc} a & 0 & 0 & 0 & d \\ & b & 2a & 0 & 0 \\ & & b & 0 & 0 \\ & & & c & 0 \\ & & & & c + 3a \end{array} \qquad \text{(M6)}$$

The loss of two degrees of freedom is accidental (or designed) rather than symmetry-determined.

Trigonal symmetry offers another perspective on redundancy. An $M(A\text{–}A)_3$ complex, with D_3 symmetry, is a convenient example. The d orbitals split into an a_1 and two e representations. Symmetry-adapted d orbitals can be constructed at the outset (the d_{z^2} orbital (a_1) is aligned on the C_3 axis). There are thus just three distinct eigenvalues. An additional parameter is necessary, however, to express the mixing between the e-symmetry orbitals. The magnitude of this parameter depends on the extent of the deviation from an octahedral arrangement of the A groups. Subtracting one for the arbitrary baseline of the diagonal elements, there are three ligand field parameters necessary to express all energy differences. These have been expressed in a number of different ways.

If the tetragonal $\{d_{xy}, d_{xz}, d_{yz}, d_{x^2-y^2}, d_{z^2}\}$ basis is used, the ligand field potential matrix for $M(A\text{–}A)_3$ has the general form

$$\begin{matrix} a & d & d & 0 & 2c \\ & a & d & -\sqrt{3}c & -c \\ & & a & \sqrt{3}c & -c \\ & & & b & 0 \\ & & & & b \end{matrix} \qquad \text{(M7)}$$

The four degrees of freedom (three spectroscopically independent) are thus apparent. If the matrix elements are put into correspondence with AOM parameters and angular variables, several situations can arise. The same angular variables apply as for the previously discussed A–A bidentate ligand: the Cartesian bite angle α, the Cartesian twist angle β, and the π-orientation angle ψ. AOM parameters $e_{\sigma A}$ and $e_{\pi A}$ bring the total to five, which cannot be gleaned from the three degrees of freedom optimally determinable from spectroscopy.

If the A group is a saturated amine only three parameters remain: $e_{\sigma A}$, α, and β. It should be possible to determine all three from the ligand field potential matrix, or directly from the spectroscopic fit. Our own experience with the ethylenediamine complex, $[Cr(en)_3]^{3+}$, revealed another problem [10]: the effects of α and β on the ligand field potential matrix elements and on the calculated band positions, while not identical, are similar enough that they cannot be practically disentangled. It was, however, possible to determine α and β in $[Cr(en)_3]^{3+}$ by using circular dichroism spectra (see Sect. 7).

Assuming that the elements of the ligand field potential matrix can be determined from the spectral information, the diagonal elements are known only to within an additive constant, while the signs of the off-diagonal elements cannot be determined. It may seem that to determine the values of all AOM parameters one must invariably have a fixed value for one, in order to fix the additive constant. In fact, this need not be so, since the AOM parameters in the diagonal are also involved with off-diagonal elements, whose magnitudes are found exactly (as far as the model is concerned) in the fitting procedure. If the geometrical factors are known (for example, by means of an X-ray crystal structure), then only the AOM parameters are left to be extracted from the ligand field potential matrix. That this works in principle can be easily tested by choosing $e_{\sigma A}$ and $e_{\pi A}$ arbitrarily, and fixing α, β, and ψ for an $M(A\text{–}A)_3$ complex (matrix M7) to generate a spectrum artificially. The ligand field potential matrix can be recovered, except for the arbitrary constant on the diagonal, by fitting the spectrum. Both $e_{\sigma A}$ and $e_{\pi A}$ can then easily be recovered from the ligand field potential matrix, no matter what constant is added to the diagonal.

It is thus *not* generally true that AOM parameters can only be determined if one of them is fixed, nor are their magnitudes arbitrary. In O_h (matrix M2) and

D_{4h} (matrix M3) symmetry it does happen that one parameter must be fixed, but that is because there are no off-diagonal elements in the ligand field potential matrix.

3 Angular Geometry in $[Cr(NH_3)_5OH]^{2+}$

The most important factor determining the sharp line splittings in the hydroxopentaamminechromium(III) complex is the orientation of the *p* orbital on the oxygen atom. This is because hydroxide is the only π-bonding ligand in the complex, and in the limit of 90° bond angles, only π-bonding affects the energies of the t_{2g} orbitals. It is reasonable to assume that the atoms in the CrN_5O skeleton do lie on the Cartesian axes, because the small deviations expected will affect the spectrum relatively little. M–O–H angles tend to be near 120° degrees, so it can be presumed that the oxygen is sp^2 hybridized, and the remaining *p* orbital is perpendicular to the plane defined by the Cr–O–H atoms. The position of the hydrogen thus dictates whether the *p* orbital overlaps with just one metal d orbital (d_{xz} or d_{yz}) or with both. A particular N–Cr–O–H dihedral angle (ψ) can vary uniquely from 0° (hydrogen eclipses a nitrogen) to 45° (hydrogen is midway between two nitrogens). The ligand field potential matrix has the form

$$\begin{matrix} a & 0 & 0 & 0 & 0 \\ & b & e & 0 & 0 \\ & & b & 0 & 0 \\ & & & c & 0 \\ & & & & d \end{matrix} \qquad \text{(M8)}$$

A maximum of four parameters can be retrieved from the spectrum. This allows $e_{\sigma N}$, $e_{\sigma O}$, $e_{\pi O}$, and ψ to be determined, in principle. Figure 2 shows the calculated variation of the sharp line ($^4A_{2g} \rightarrow {}^2E_g, {}^2T_{1g}, {}^2T_{2g}$) transition energies as a function of ψ [5]. The periodicity in Fig. 2 occurs because of the fourfold potential around the N–Cr–O axis. Two of the lines are extremely sensitive to the dihedral angle. The potential variation by several hundred cm^{-1} in the splittings with neighboring bands stands in contrast to line widths of 5 to 10 cm^{-1} in spectra at 12 K. The splittings are much less sensitive to the ligand field properties than to the geometry. In complexes with approximate tetragonal symmetry the splittings are particularly insensitive to differences in Dq ($= 0.3e_\sigma - 0.4e_\pi$), though more sensitive to additive combinations such as $e_\sigma + e_\pi$ [12].

It is difficult to formulate an expectation for the value of ψ. Molecular mechanics calculations indicate that the strain energy changes little between 0° and 45°. The strong variation of transition energies with dihedral angle would lead one to presume that the value could be quite accurately determined from

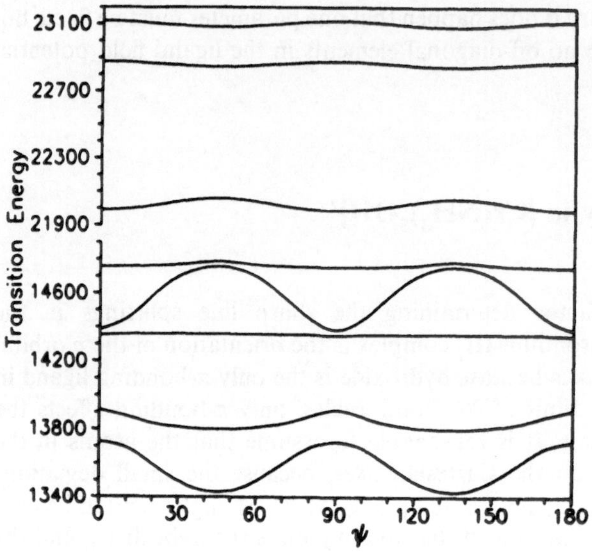

Fig. 2. Calculated variation of the $t_{2g} \rightarrow t_{2g}$ sharp-line transition energies in $[Cr(NH_3)_5OH]^{2+}$ as a function of the dihedral N–Cr–O–H angle ψ [5]. Other parameters (in cm^{-1}): $e_{\sigma N} = 7180$, $e_{\pi O} = 2900$, $B = 540$, $C = 3200$, $\alpha_T = 0$, $\zeta = 270$. The upper three curves refer to the octahedral $^4A_{2g} \rightarrow {}^2T_{2g}$ transition; the lower five refer to the $^4A_{2g} \rightarrow {}^2E_g$ and $^2T_{1g}$ transitions, which are mixed so that the parentage changes markedly with the angle ψ

the spectrum. But here a problem that plagues the field crops up. The assignment of the electronic peaks in a spectrum with many vibronic components is far from straightforward. There are two published assignments of the $^4A_{2g} \rightarrow \{^2E_g, {}^2T_{1g}\}$ lines in the sharp-line spectrum, both with plausible justification. If Güdel's assignment [13], in which the splitting of the first two bands is 90 cm^{-1}, is correct, the dihedral angle works out to be $5.2 \pm 0.2°$. If our own assignment [5], in which the splitting of the first two bands is 630 cm^{-1}, is correct, then the dihedral angle is $20.6 \pm 0.5°$. The error margins attributed to these values are those propagated from the presumed uncertainties in the peak positions.

The actual angle is not yet known, and it will require a very careful X-ray structure determination to fix the hydrogen position. It is unlikely that there is much variation in this position in the crystal, because if there were, the electronic lines that depend on the dihedral angle would be very broad.

4 Beyond the First Coordination Sphere

Thus far, only the potential from the six coordinated atoms on the metal d orbitals has been discussed. These atoms must produce, by far, the largest perturbations on the d orbital energies. But other parts of the ligand, counterions, and even the other complexes in the crystal may also contribute. Very often these contributions arise from an arrangement of lower symmetry than that of the coordination skeleton. This may, in some circumstances, lead to larger effects on the line splittings from the periphery than from the first coordination sphere.

Salts of $[Cr(CN)_6]^{3-}$ provide a useful example. The ^2E_g splittings, which would be zero for perfectly octahedral symmetry, are commonly between 50 and 100 cm^{-1} [14, 15], larger than in many complexes of lower symmetry. The geometry of the potassium salt is known from X-ray structure determinations [16, 17]. The site symmetry of the chromium is only maximally C_i, but the slight deviations from octahedral symmetry are not large enough to cause such large splittings. Figure 3 shows the disposition of the eight nearest potassium ions around the complex. These K^+ ions form a distorted cube, and another set of six ions further away form a distorted octahedron, the axes of which are approximately aligned with those of the distorted cube. However, the axes of the $[Cr(CN)_6]^{3-}$ octahedron are not similarly aligned, as is seen in Fig. 3. The counterions therefore constitute a low symmetry perturbation on the nearly octahedral *d*-electron energy levels of the Cr(III).

Another factor comes into play here. In the first coordination sphere, deviations in metal-ligand bond distances are normally small enough to be neglected. The nearest 14 K^+ ions lie at seven separate distances, from 4.3 to 6.0 Å. The variation in the magnitude of the perturbation with distance must be considered, and it is not straightforward. Classical crystal field theory shows an inverse fifth power dependence of Dq on distance from the metal, but this result is specific to octahedral symmetry, in which lower order dependencies drop out. For individual perturbers we have proposed a dependence of the form $aR^{-3} + bR^{-5}$ for both e_σ and e_π [7]. For counterions we also suggest that the π contribution and the R^{-5} part of the σ contribution can be neglected.

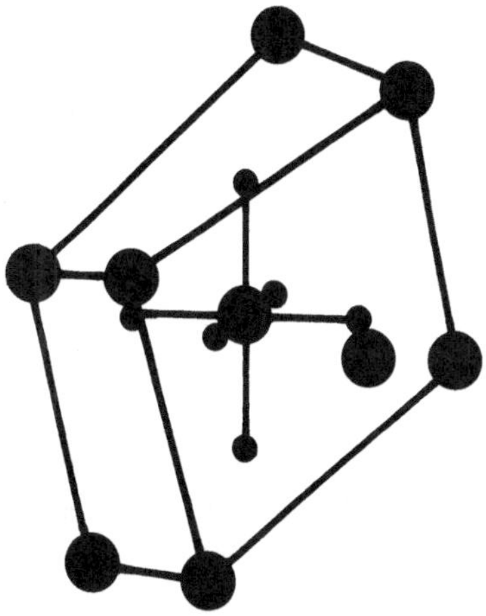

Fig. 3. Orientation of $[Cr(CN)_6]^{3-}$ within the eight nearest K^+ ions. The nitrogens are omitted. Atom positions are from Ref. 16

All eight sharp line peaks have been measured and assigned for $K_3[Cr(CN)_6]$ [7, 18, 19]. Without including a cation field, the optimized fit to the spectrum was poor. The calculated ^2E_g splitting was 11 cm^{-1} (experimental value 53 cm^{-1}), and the fit for several other lines was even worse. With the field from the nearest 14 cations introduced as described above, the fit improved markedly, though the ^2E_g splitting was still just 21 cm^{-1}. The optimum value of e_σ for K^+ was -320 cm^{-1} for the nearest ion, decreasing as R^{-3} (by assumption) for those further away. The negative sign is appropriate for the perturbation from a positive charge, and appeared in the optimization even when positive values were introduced as initial guesses. It is possible that the fit would improve further if more counterions, and possibly neighboring $[Cr(CN)_6]^{3-}$ ions, were included in the field.

Just this approach was taken in the analysis of the spectrum of ruby [20], which is Cr^{3+} doped in Al_2O_3. The chromium ion lies on a site of C_3 symmetry [21], so the ligand field potential matrix is similar to M7, and has three recoverable degrees of freedom. We represented the ligand field in terms of e_σ and e_π for oxide and e_σ for Al^{3+} ion. In an ionic crystal there is no obvious demarcation for the coordination environment, and how large an environment one chooses to represent the ligand field may seem arbitrary. If one assumes, as in the preceding example, an R^{-3} dependence for all three AOM parameters, the reduction in the ligand field from individual ions with distance would be partially compensated for by the R^2 increase in the number of ions with distance, so that the total contribution to the ligand field would fall off with distance from the Cr^{3+} as $1/R$. This is not as rapid as one would like if one is to define a cutoff distance for the coordination environment.

Furthermore, it appears that the oxide ion AOM parameters do fall off as R^{-3}, but $e_{\sigma Al}$ is better represented as decreasing with R^{-5}. This was determined by initially allowing both parameters to vary with distance as $R^{-3} + q \cdot R^{-5}$. The values of q that resulted indicated that the contribution from the R^{-5} term was small for $e_{\sigma O}$ and $e_{\pi O}$, and likewise the R^{-3} term was relatively unimportant for $e_{\sigma Al}$. This is not unreasonable, because the nearest Al^{3+} ions form a much more symmetric (that is, nearly cubic) arrangement about Cr^{3+} than do the nearest oxide ions. As in the octahedron, the R^{-3} dependence tends to drop out.

What we found was that the larger the coordination environment we included in the ligand field, the better the optimized fit. The largest set we used included all ions within 15 Å, balanced for charge, a total of 1355 ions. From the values of the AOM parameters obtained for the nearest oxide ions ($e_{\sigma O} = 8870$ cm^{-1}, $e_{\pi O} = 850$ cm^{-1}), it is characterized as a moderate σ-donor and a weak π-donor [20]. The latter behavior was unexpected, since ligands like hydroxide and carboxylate are fairly strong π-donors. In the aluminum oxide crystal, however, oxide is probably forced to use its valence electrons in a σ-interaction with a tetrahedral array of Al^{3+} ions around it. The value of -9200 cm^{-1} for e_σ of the nearest Al^{3+} ions has the right sign, but seems quite large, even for a + 3 ion, considering that they are approximately 50% further from the chromium than the nearest oxide ions.

Table 1. Some calculated and observed transition energies for Cr^{3+}: Al_2O_3 (cm^{-1})

Excited state	Macfarlane[a]	Veremeichik[b]	Our laboratory[c]	Exptl[d]
$^4A_{2g}$ splitting	0.31		0.38	0.38
^2E_g	14054	14352	14415	14418
	14077		14437	14447
splitting	23		22	29
$^2T_{1g}$	14611	15012	15006	14957
	14807	15112	15070	15168
	14817		15109	15190
$^2T_{2g}$	21590	21062	21072	20993
	21614		21109	21068
	21887	21393	21473	21357

[a] Macfarlane RM (1963) J Chem Phys 39: 3118. [b] Veremeichik TF, Grechushnikov BN, Kalinkina, IN (1986) Zh Prikl Spektrosk (English Translation) 44: 620. [c] Lee K-W, Hoggard PE (1990) Inorg Chem 29: 850; consult this reference for calculated and observed values for transitions to other excited states. [d] Fairbank, WM Jr, Klauminzer GK, Schawlow AL (1967) Phys Rev 155: 296.

The resulting fit, with the nearest 1355 ions, improved considerably on earlier efforts, and was especially good for the zero field splitting (within $0.01\ cm^{-1}$) and the ^2E components (within $10\ cm^{-1}$). Table 1 presents a comparison of calculated and experimental values for the transitions within the $(t_{2g})^3$ configuration.

5 The Bipyridine Ligand

The spectrum of $[Cr(bpy)_3]^{3+}$, bpy = 2,2′-bipyridine, has proved to be quite controversial. Hauser et al. [22] attributed a value of $20\ cm^{-1}$ to the ^2E_g splitting in the PF_6^- salt, and considered that value to be in excellent agreement with ligand field theory, which predicts a very small splitting in a complex with approximate D_3 symmetry, since the twofold degeneracy of an octahedral E state is not lifted upon descent to D_3. Including the spin part, splitting does occur, but it would still be small for metal ions like chromium(III) with relatively limited spin-orbit coupling.

Working with the perchlorate salt in our laboratory, however, we could find no evidence of intervals attributable to a small ^2E_g splitting. Instead, a line $165\ cm^{-1}$ above the lowest energy electronic line was assigned to the second component of the ^2E_g state. Classically, this splitting is much too large. Bipyridine, however, has two properties that are not accounted for in classical ligand field treatments: the π-interaction with the metal is anisotropic, and the nitrogen atoms are connected by conjugated double bonds. Anisotropic π-bonding (perpendicular to the bpy ring) is expected in chelating ligands, and can be readily treated within the angular overlap model framework. The conjugation of the nitrogens implies that it is not accurate to treat each nitrogen separately, as is normally done in an AOM treatment. A better view of the interaction is that the nitrogens are "phase-coupled" [23], such that a molecular orbital (MO) in which they are in phase and an MO in which they are out-of-phase interact with the metal with different effects on the d-orbital energies.

There is some controversy in the literature over how best to treat phase coupling [23–27]. Our procedure is to abandon the 3 × 3 AOM matrix for individual ligating atoms. This matrix represents interaction in a sigma (metal-ligand bond) direction, and two perpendicular π directions, and has the form M9 for an anisotropically π-bonding ligand.

$$\begin{matrix} e_\sigma & 0 & 0 \\ 0 & e_\pi & 0 \\ 0 & 0 & 0 \end{matrix} \qquad \text{(M9)}$$

There would be a third diagonal term equal to e_π if the ligand were isotropic. This matrix is pre- and postmultiplied by angular factors necessary to rotate the ligand orbitals into position to overlap with the appropriate metal d orbitals (d_{z^2} and d_{yz}) and back again [2]. For phase-coupled ligands with two coordinating atoms, those atoms are treated in pairs, and the AOM matrix expands to 6 × 6 with an off-diagonal element ($e_{\pi\pi}$) affecting the π-interaction.

$$\begin{matrix} e_\sigma & 0 & 0 & 0 & 0 & 0 \\ 0 & e_\sigma & 0 & 0 & e_{\pi\pi} & 0 \\ 0 & 0 & 0 & 0 & 0 & 0 \\ 0 & 0 & 0 & e_\sigma & 0 & 0 \\ 0 & e_{\pi\pi} & 0 & 0 & e_\pi & 0 \\ 0 & 0 & 0 & 0 & 0 & 0 \end{matrix} \qquad \text{(M10)}$$

Anisotropic π-bonding and phase coupling can each lead to splittings far in excess of what is predicted by classical ligand field theory. When the AOM matrix M10 is used, the ligand field potential matrix for $M(A\text{–}A)_3$ assumes the form

$$\begin{matrix} a & d & e & 0 & 2f \\ & a & e & -\sqrt{3f} & -f \\ & & b & \sqrt{f} & -f \\ & & & c & 0 \\ & & & & c \end{matrix} \qquad \text{(M11)}$$

Thus, up to five parameters can be extracted. A fit to the experimental spectrum was undertaken using the geometry from the room temperature X-ray crystal structure [28] and the parameter set e_σ, e_π, and $e_{\pi\pi}$. The π parameters obtained, $e_\pi = -2283\ \text{cm}^{-1}$, $e_{\pi\pi} = 342\ \text{cm}^{-1}$, can be used to express the π-interaction of one bipyridine in the form of the 2 × 2 matrix [25]

$$\begin{matrix} e_\pi & e_{\pi\pi} \\ e_{\pi\pi} & e_\pi \end{matrix} = \begin{matrix} -2283 & 342 \\ 342 & -2283 \end{matrix} \qquad \text{(M12)}$$

This matrix has the eigenvalues $-2625\ cm^{-1}$ for the out-of-phase combination, and $-1941\ cm^{-1}$ for the in-phase combination of nitrogen π orbitals. A MOPAC [29] calculation done subsequently indicated that the bipyridine LUMO is primarily nitrogen-centered and out-of-phase. It should be noted, though, that the calculations show more than two low-lying empty bipyridine orbitals with substantial contributions from the nitrogens. It is reasonable, however, that the lowest-lying of these causes the greatest stabilization of the metal d orbitals, in accord with our results.

6 Orbital Occupancies

Another use to which these calculations can be put is in evaluating the orbital occupancy in ground and excited states. For chromium(III) the ground state in a very wide variety of molecules can be expressed as almost exactly one electron in each of the three t_{2g} orbitals. The orbital composition of the lowest excited doublet state is the same as that of the ground state in O_h symmetry. But distortions from O_h, particularly tetragonal distortions, can lead to sizable changes in the orbital occupancy of the lowest doublet, which may have consequences for photochemical behavior [30, 31]. For example, in *trans*-$[Cr(NH_3)_4F_2]^+$, while the orbital composition of the ground state ($^4B_{1g}$ in D_{4h} symmetry) can be expressed as $(xy)^{1.00}(xz)^{1.00}(yz)^{1.00}$, the first excited state (^2E_g, which is derived from the octahedral $^2T_{1g}$) is calculated to have the composition $(xy)^{1.72}(xz)^{0.63}(yz)^{0.63}$ [32].

The best chances of creating, for the lowest doublet state, a substantial vacancy in one of the t_{2g} orbitals is to construct a complex placing the maximum degree of π-withdrawal on two of the t_{2g} orbitals and, if possible, some degree of π-donation on the other [33]. This is probably best accomplished by means of π-anisotropic ligands, that is, ligands using just one π-symmetry orbital to interact with the metal. Chelating ligands provide the most common examples of π-anisotropy.

Chelate complexes, however, can have holes in the electron distribution merely because of the extent of the distortion of the coordination sphere from octahedral. The best way to see this is not from the occupation numbers themselves, but from the electron density function

$$\rho(x, y, z) = \sum_{i=1}^{5} n_i[\xi_i(x, y, z)]^2 \qquad (4)$$

Here ρ is the point electron density and n_i is the occupation number of the d-orbital eigenfunction ξ_i. The first excited state (^2E) of a $D_3[Cr(A\text{–}A)_3]^{3+}$ complex with no π-interaction and Cartesian bite and twist angles $\alpha = 9°$ and $\beta = 9°$ is calculated to have the electron density on the two faces perpendicular to the C_3 axis reduced to one quarter of the electron density that would be present in octahedral geometry [34]. That is a rather large, though not unheard of, distortion. The ground state is calculated to have only small differences in electron density compared to octahedral geometry.

An example in which π effects play the major role is *trans*-$[Cr(NH_3)_2(L\text{–}L)_2]^{3+/1-}$, in which the L–L ligand is strongly π-electron withdrawing. The ammonias can be leaving groups under irradiation. In this complex the regions of low electron density in the lowest excited state are the two edges in the xy-plane not occupied by an L–L bridge. With AOM parameters $e_{\sigma N} = 6500$, $e_{\sigma L} = 6500$, and $e_{\pi L} = -1500\ cm^{-1}$, the densities on the edges described are calculated to be 40% of what they would be if the field were octahedral, while in the ground state all the edge densities are approximately the same [34]. This can be thought of as due to withdrawal of electron density in the xz and yz planes in the excited state by the anisotropic π-symmetry orbitals on the L ligands, while the xy plane is unaffected. If an associative photosubstitution pathway exists, involving nucleophilic attack on the doublet excited state, this would lead to the prediction that the rate would increase the greater the π-withdrawing power of the L–L ligand.

7 Rotational Strengths and Angular Geometry

Metal complexes with bidentate saturated amines ought to be good candidates for extraction of geometric information from the spectrum. Tris complexes with D_3 symmetry yield a ligand field potential matrix of the form M7. There are three recoverable parameters, while e_σ plus the bite (α) and twist (β) angles are all that is necessary to describe the ligand field from the MN_6 skeleton. The problem is that α and β have fairly similar effects on the matrix. In ideal cases, for example when the spectrum is simulated, all three parameters can be readily extracted from the spectrum. Given a real spectrum, the error margins for α and β are often as large as the angles themselves.

$M(A\text{–}A)_3$ complexes are optically active, and the problem can be remedied if a circular dichroism spectrum of one enantiomer can be measured and the sharp electronic lines identified. The bite and twist angles have very different effects on the rotational strengths. The twist angle has a marked effect that is quite plausible when one considers that a 60° twist causes conversion to the opposite enantiomer. Sign changes in the rotational strength can also occur at twist angles near $\beta = 0°$.

The rotational strength of an electronic transition depends on both the electric and magnetic transition dipole moments [35]:

$$R(0 \rightarrow j) = \mathrm{Im}\{\langle 0|\mathbf{E}_x|j\rangle\langle j|\mathbf{M}_x|0\rangle + \langle 0|\mathbf{E}_y|j\rangle\langle j|\mathbf{M}_y|0\rangle + \langle 0|\mathbf{E}_z|j\rangle\langle j|\mathbf{M}_z|0\rangle\} \quad (5)$$

E and **M** are the electric and magnetic dipole moment operators, respectively, and only the imaginary part of the expression is taken to evaluate the rotational strength of a transition between the ground state and an excited state (ψ_j). Evaluation of the rotational strength requires the 5×5 matrices of these operators within the d orbital basis. These are well known for the magnetic dipole operator [36], but the electric dipole elements, $\langle d_i | E_\alpha | d_j \rangle$, are all zero because the operator is odd. In dissymmetric environments, however, the d orbitals can mix with odd-parity orbitals on the metal or ligands. We have calculated rotational strengths based on the mixing of the d orbitals with the valence p orbitals on the metal ion. The mixing under the influence of the external field can be evaluated with the same AOM framework used to find the ligand field potential matrix $\langle d_i | V | d_j \rangle$ [10]. The analogous $\langle p_i | V | d_j \rangle$ matrix can be derived for any number of perturbing atoms at their exact positions. For a D_3 $M(A\text{–}A)_3$ complex oriented with the trigonal axis along the (1, 1, 1) direction it has a particularly simple form:

$$\begin{matrix} -2a & 2a & 0 & -a & -\sqrt{3a} \\ 2a & 0 & -2a & -a & \sqrt{3a} \\ 0 & -2a & 2a & 2a & 0 \end{matrix} \qquad \text{(M13)}$$

The rows are in the order p_x, p_y, p_z, and the columns in the order d_{xy}, d_{xz}, d_{yz}, $d_{x^2-y^2}$, and d_{z^2}. The electric dipole matrix elements between p and d orbitals are easily calculated [10], and inserted into the perturbation expression to find the matrix elements between the d orbitals to within a multiplicative constant.

An important question is where the dissymmetric perturbation comes from. The six nitrogens, because they have D_3 symmetry around the metal, lead to nonzero rotational strengths by themselves. But the complex would be optically active even if the nitrogens were at 90° angles, because of the chelate rings. In modeling $[Cr(en)_3]^{3+}$, this led us to also include the carbon atoms as perturbers of the metal d orbitals. Because of their distance from the metal and the lack of lone pairs for direct interaction, their influence on the d orbital energies is small. But because their positions are much further from octahedral symmetry, they have probably a greater influence on rotational strengths than the nitrogens do. The flexible ethylenediamine rings preferentially adopt one of the two stable conformations, *lel* and *ob* (*lel* is the conformation in which the ethylenediamine C–C bond axes are nearly parallel to the C_3 axis of the complex). This can have a significant effect on the spectrum observed.

The low-temperature CD spectrum of crystalline Λ-$[Cr(en)_3]Cl_3$ was measured by Geiser and Güdel [37]. The sign of $\Delta\varepsilon$ was found to be positive for the first four of the ${}^4A_{2g} \rightarrow \{{}^2E_g, {}^2T_{1g}\}$ transitions, while the fifth was negative, but very weak (and uncertain). The room temperature solution CD spectrum differs markedly [38, 39]. The line widths were such that only three lines are observed. The first, comprising the two ${}^2E_g$ components, had a positive $\Delta\varepsilon$. The second was negative, and the third (the fourth and fifth components) was positive.

We used the room temperature crystal structure [40] to fit the transition energies and rotational strengths of Λ-$[Cr(en)_3]Cl_3$. The calculated spectrum resulting from the best fit parameters had a positive $\Delta\varepsilon$ for all five doublet peaks. This can be considered reasonably good agreement with experiment. In analyzing the solution spectrum we attempted to find the Cartesian bite and twist angles, α and β, by means of the fitting procedure. It was expected that they would be considerably different than in the solid, for which the average value of α is about 4° (the rings are not equivalent) and the average value of β about 2°. It would seem that a substantial change would be necessary in order to cause the difference in the two CD spectra.

Another possibility, however, was that conformational differences in the rings were responsible for most of the spectral differences. The ring conformation in the room-temperature solid is lel_2ob [40]. One might expect either the *lel* or *ob* conformation to be energetically favored and ascribe the mixed conformation to crystal packing forces. A more symmetric lel_3 or ob_3 conformation in solution could then be responsible for the differences in the CD spectra.

Attempts to fit the spectrum with either of the more symmetric conformations yielded unsatisfactory results for the signs of $\Delta\varepsilon$, even though this was weighted heavily in the fitting procedure. Instead, the lel_2ob conformation yielded the best fit, and matched the signs of all three doublet bands in the CD spectrum [10]. The only important difference between the geometry derived for the complex in solution and the solid state geometry was that β was fixed by the optimization process at 1.5° in solution, approximately 0.5° smaller than the average value in the crystal.

It is remarkable that this small difference in the twist angle could cause a major difference in the CD spectra. Yet our modeling of the effects of β on the rotational strengths of the doublet lines and on the first spin-allowed band show that exactly this behavior is to be expected. Thus we conclude that very similar geometries exist in solution and in the solid state, and the controversy in the literature over the proper assignment of the CD spectrum [37–39], may be settled, since both sets of assignments appear to be correct.

8 An Overview of AOM Parameters

AOM parameters have now been derived for a number of ligands using the actual geometry of the complex rather than a more symmetric approximate geometry. As these data accumulate, there are two types of consistency one would like to see if the model is a good one. First, a given ligand, such as ethylenediamine, should have consistent values for its AOM parameters in different complexes – either the parameters should not vary significantly, or they should vary in a reasonable and predictable way that depends on the other ligands in the complex. Second, one would like a given ligating group, such as a primary amine, to yield consistent values for its AOM parameters when incorporated in different ligands that complex with different geometries.

This is the issue of transferability, which is one of the reasons that the Angular Overlap Model was conceived. Except for Dq, and sometimes parameters that express differences in Dq, ligand field parameters tend not to be transferable from one complex to another. Thus they carry little chemical information. AOM parameters become useful to the extent that consistency can be demonstrated. Then trends within series of related ligands will have meaning, and comparisons with nonspectroscopic properties can be undertaken.

Table 2 collects AOM parameter values that have been assigned in spectroscopic studies in which aspects of the geometry were not treated as variables. A fixed geometry, usually from an X-ray crystal structure, was used to express the angular dependence of the ligand field. Studies to evaluate consistency for individual ligands have yet to be undertaken. One interesting comparison, though, is between the related ligands oxalate and malonate. Oxalate coordinates with a Cartesian bite angle of about 4°, which becomes approximately 0° for malonate. From Table 2 it can be seen that $e_{\sigma O}$ is virtually the same for both. The difference in $e_{\pi O}$ is sizable, but the two ligands are nevertheless similar in their π-interaction properties.

Table 2. AOM parameters (cm^{-1}) derived from chromium(III) complexes with known geometry

AOM parameter	Ligand	Value	Complex	Ref.
$e_{\sigma C}$	CN^-	9990	$K_3[Cr(CN)]_6$	[7]
$e_{\pi C}$	CN^-	600	$K_3[Cr(CN)]_6$	[7]
$e_{\sigma N}$	en	7590	$[Cr(en)_3]Cl_3$	[10]
	tacn	7610	$[Cr(tacn)_2]Cl_3$ [b]	[41]
	$tcta^{3-}$	7050	[Cr(tcta)]	[41]
	dpt	(7370)[a]	$[Cr(dpt)(glygly)]ClO_4$	[6]
	$glygly^{2-}$ (peptide)	(7430)[a]	$[Cr(dpt)(glygly)]ClO_4$	[6]
	bpy	5650	$[Cr(bpy)_3](ClO_4)_3 \cdot HbpyClO_4$	[25]
	phen	5380	$[Cr(phen)_3](ClO_4)_3$	[42]
	terpy	7230	$[Cr(terpy)_2](ClO_4)_3$	[42]
$e_{\pi N}$	$glygly^{2-}$	(500)[a]	$[Cr(dpt)(glygly)]ClO_4$	[6]
	bpy	−2280	$[Cr(bpy)_3](ClO_4)_3 \cdot HbpyClO_4$	[25]
	phen	−2410	$[Cr(phen)_3](ClO_4)_3$	[42]
	terpy	−140	$[Cr(terpy)_2](ClO_4)_3$	[42]
$e_{\sigma O}$	O^{2-}	8870	ruby	[20]
	$tcta^{3-}$	7910	[Cr(tcta)]	[41]
	ox^{2-}	7110	$K_3[Cr(ox)_3]$	[42]
	mal^{2-}	7150	$K_3[Cr(mal)_3]$	[42]
	$glygly^{2-}$	(7330)[a]	$[Cr(dpt)(glygly)]ClO_4$	[6]
$e_{\pi O}$	O^{2-}	850	ruby	[20]
	$tcta^{3-}$	2040	[Cr(tcta)]	[41]
	ox^{2-}	1440	$K_3[Cr(ox)_3]$	[42]
	mal^{2-}	1880	$K_3[Cr(mal)_3]$	[42]
	$glygly^{2-}$	(1240)[a]	$[Cr(dpt)(glygly)]ClO_4$	[6]

ligand abbreviations: en = ethylenediamine, ox^{2-} = oxalate, mal^{2-} = malonate, tacn = 1,4,7-triazacyclononane, $tcta^{3-}$ = 1,4,7-triazacyclononane-*N*,*N*′,*N*″-triacetate, bpy = 2,2′-bipyridine, phen = *o*-phenanthroline, terpy = 2,2′,2″-terpyridine, dpt = di(3-aminopropyl)amine, $H_2glygly$ = glycylglycine.

[a] Geometry based on molecular mechanics calculation. [b] Geometry of $[Fe(tacn)_2]Cl_3$ [56]

9 An Evaluation

9.1 Experimental Obstacles

The determination of bond angles or ligand field parameters from electronic spectra, including especially the sharp electronic lines, can be far from straightforward. The problems to be discussed below should not obscure one point, however: it is better to take the metal-ligand geometry (actual or approximated, by molecular mechanics calculations for example) into account than to ignore it and assume some ideal geometry. The most common assumption is that the bond angles are 90° (orthoaxiality) [43]. This can be quite satisfactory for deriving qualitative information, such as which geometric isomer one has, but it is less satisfactory for quantitative information, namely ligand field parameters.

For example, for a $[Cr(A\text{–}A)_3]^{3+}$ complex with AOM parameter values $e_{\sigma A} = 7000\ cm^{-1}$ and $e_{\pi A} = 0$, the first band maximum should be found at $21\,000\ cm^{-1}$ if the Cartesian bite and twist angles are 0°, that is, for an octahedral CrA_6 skeleton. If the Cartesian bite angle is 6°, however, the model predicts the first band maximum to fall at about $20\,000\ cm^{-1}$, which with the traditional approximate ligand field treatments would be taken as the value of 10Dq (see Table 3). One application of AOM parameter values is to assess bond strengths [44], which are related to $e_\sigma + ze_\pi$, where the value of z depends on the d-electron configuration. Such correlations should be more valuable when the AOM parameters are determined so as to factor out the influence on the spectrum of the angular geometry.

Among the problems in applying these techniques to determine geometry or AOM parameter values, or both, is that it is quite often difficult to identify the sharp electronic lines in the spectrum. There are several factors at work. Vibronic lines associated with one sharp electronic line can often mask other electronic lines. When the metal is at a site that is nearly centrosymmetric, the electronic origins can be extremely weak.

Some of the features that have been used to assign lines in an absorption or excitation spectrum as electronic or vibrational are listed below. What follows

Table 3. Calculated $^4A_2 \rightarrow {}^4T_2$ transition energies and inferred AOM parameter values for a $Cr(A\text{–}A)_3$ complex with different Cartesian bite angles[a]

α (deg)	$E(d_\sigma)$[b]	$E(^4T_2)$[c]	$e_{\sigma A}$
0	21000	21000	7000
3	20828	20741 (8) 20756 (4)	6915
6	20319	19973 (8) 20037 (4)	6665

[a] In cm^{-1}; input values: $e_{\sigma A} = 7000$, B = 650, C = 3200, $\zeta = 0$. [b] One-electron energy of $d_{x^2-y^2}$ and d_{z^2} orbitals. [c] Degeneracies in parentheses.

assumes that the luminescence is sharp-line, arising from the lowest excited state of the same configuration as the ground state.

1. Intensity – the presumption is that very intense lines are electronic origins.
2. Coincidence of an interval in the absorption or excitation spectrum with a line in the luminescence spectrum (or perhaps the IR or Raman spectrum) – this would be evidence that the lower energy peak is an electronic origin and the higher energy peak a vibronic satellite; if a line cannot be associated with a known vibronic interval based on any lower energy electronic origin, it is evidence that the line is itself an electronic origin.
3. Similar, overlapping sequences of vibronic satellites – it is presumed that the electronic origins are those lines on which the most consistent sets of vibronic satellites can be constructed from among the observed spectral lines.

Assignments of electronic origins can also be colored by expectations based on the spectra of related compounds. One must be extremely cautious here not to make a false comparison. As discussed in Sect. 5, a tris complex of 2,2′-bipyridine is not similar to a tris complex of ethylenediamine, even though the skeleton is MN_6 in both cases, and the skeletal geometry may be comparable.

There are problems with each of these criteria. Intensities of vibronic satellites *are* often lower than the intensity of their electronic origin, but when there are several electronic lines of different intensity, it is easy for some to be weaker than nearby vibronic satellites belonging to other origins. When the complex approaches centrosymmetry, the electronic origins become more forbidden, and can easily be dwarfed by their own satellites.

Figures 4 and 5 illustrate this point. The electronic lines in the facial tris(glycinato) complex are all more intense than the associated vibronic structure. This complex is fairly distant from centrosymmetry, because each amine group is nearly opposite a carboxylate. In the spectrum of the tris(ethylenediamine) complex, however, vibronic structure is more prominent. Though there is no actual inversion center, amines are nearly opposite amines, and the geometry is not far from centrosymmetric.

Coincidence of an interval with a luminescence peak may be just that – a coincidence. The denser the spectrum the more likely such coincidences are. *Not* finding a match for a line in the luminescence, IR, or Raman spectrum is somewhat stronger evidence, but selection rules may account for this, and in some symmetries (D_{4h} in particular) it is actually common.

Intensities and the coincidence or noncoincidence with vibronic intervals in luminescence are frequently used jointly as the basis for the assignment of electronic origins. Because of the difficulties just mentioned, we have promoted the use of the third criterion – similar vibronic patterns built on the assigned electronic origins. This too has its drawbacks, most particularly for symmetry groups that have several one-dimensional representations. When the excited states belong to different representations, their vibronic structures will be quite different. The method works best when there is little or no symmetry. Even then, however, it is far from straightforward, and the assignments reached cannot

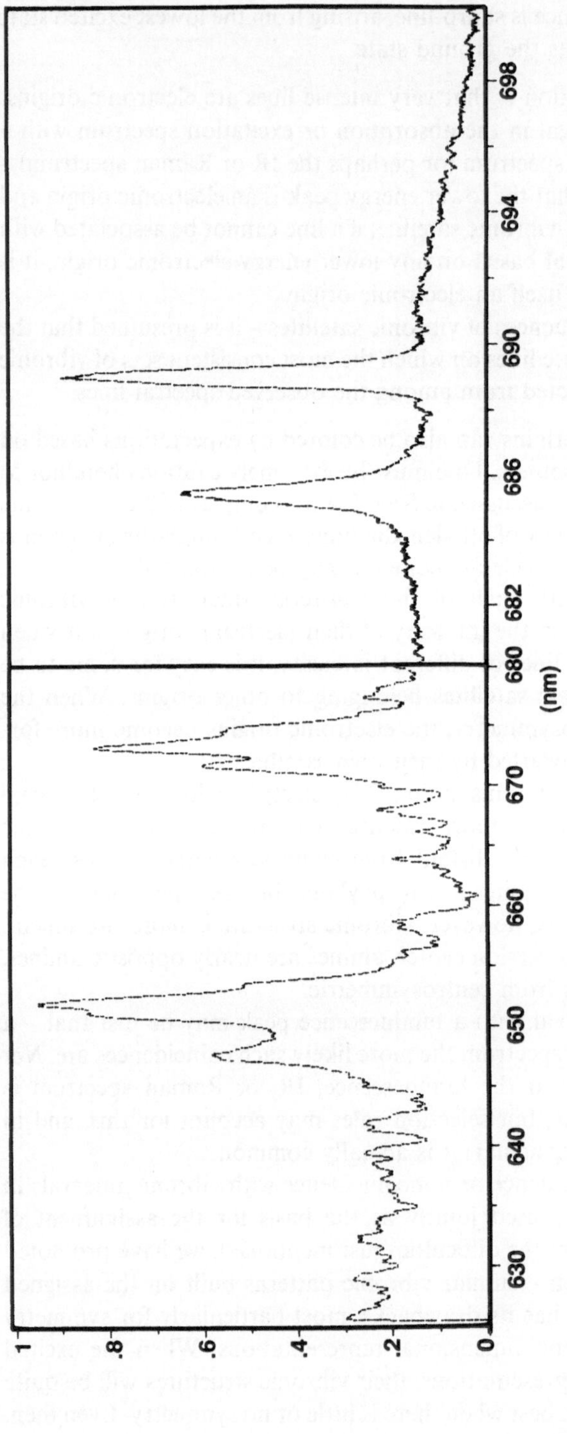

Fig. 4. Excitation spectrum of *fac*-[Cr(gly)$_3$] at 12 K in the region of the $^4A_{2g} \rightarrow {}^2E_g$ (*right*) and $^4A_{2g} \rightarrow {}^2T_{1g}$ (*left*) transitions [3]. Note abscissa scale change

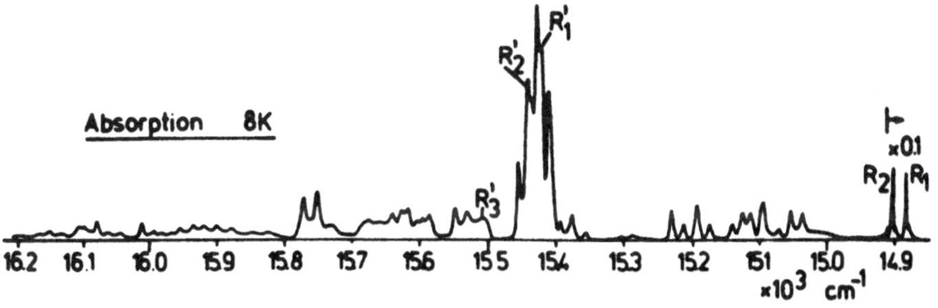

Fig. 5. Axial absorption spectrum at 8 K of $[\Lambda\text{-Cr(en)}_3][\Delta\text{-Ir(en)}_3]Cl_6 \cdot KCl \cdot 6H_2O$ [37]. The assigned $^4A_{2g} \rightarrow {}^2E_g$ electronic origins are labeled R_1 and R_2; the assigned $^4A_{2g} \rightarrow {}^2T_{1g}$ origins are labeled R'_1, R'_2, and R'_3

usually be regarded as conclusive. Groups studying the same compound but relying on different criteria can assign the electronic origins differently. The ensuing controversies cannot readily be resolved.

Another problem related to vibronic masking is that most sharp-line transitions are spin-forbidden, and easily obscured by even the tails of nearby spin-allowed bands. They commonly occur over a wide range of energies, putting greater demands on instrumentation. Low-spin, six-coordinate Os(IV) complexes with five or six chloride ligands, for example, have $t_{2g} \rightarrow t_{2g}$ transitions with energies from below 3000 to about 17000 cm^{-1} [45]. Six-coordinate V(III) complexes commonly exhibit the spin-allowed bands from $t_{2g} \rightarrow e_g$ transitions in the visible, while the $t_{2g} \rightarrow t_{2g}$ lines occur between 1000 and 1200 nm [46, 47].

Even given accurate assignments of the sharp-line electronic origins, there are still problems in deriving AOM or angular parameters from them. The mathematical fitting procedure with which the parameters are calculated results in a minimum uncertainty through the propagation of the experimental uncertainties. If the latter are properly estimated, the uncertainties in parameter values can be accurately assessed [8]. When there is a high degree of correlation between parameters, small experimental uncertainties can be transformed into large uncertainties in parameter values.

Additional uncertainty arises from the inadequacy of the model itself, both specific and nonspecific. The nonspecific refers to the failure of the angular overlap model itself to adequately represent the electronic states of the molecule, and in particular, the inconsistencies in AOM parameter values that result. This can only be assessed by compiling enough data to evaluate those inconsistencies. Specific defects are those in which the metal-ligand interaction is treated unsatisfactorily, but can be improved within the framework of the model. An example is when the π-interaction is treated as isotropic, but is actually anisotropic. Specific defects can be assessed immediately by making the proposed alteration, and this works particularly well if no additional parameters are required to make the alteration.

A classic example concerns the ^2E_g splitting in the chromium(III) pentaammines. Based on any reasonable set of ligand field or AOM parameters, the splitting should be no greater than 50 cm^{-1}. However, splittings of 200 cm^{-1} are common in this series of complexes [48]. This led us and others to propose that the observed splitting was not between the ^2E_g components at all, but between one ^2E_g component and one from the higher energy $^2T_{1g}$ group [49]. Schmidtke et al., on the other hand, proposed that the ^2E_g splittings were indeed that large, but that interelectronic repulsion was inadequately treated by using a spherical parametrization [4]. With Schmidtke's suggested correction (based on the π-electron asymmetry), the splitting could be accurately fit. The value, or even the validity, of this correction could not be established immediately, because two new parameters were introduced, although one of them did not have much influence on the splitting. Further experimental work lent support to a large ^2E_g splitting in $[Cr(NH_3)_5Cl]^{2+}$ [50], while we found that modeling the sharp-line spectra of other Cr(III) complexes with tetragonal symmetry was greatly improved by including Schmidtke's corrections as constants [5, 6].

9.2 Improving Assignments of Electronic Lines

Two experimental techniques deserve mention as possible means of resolving disputatious assignments, by discriminating against vibronic in favor of electronic lines. One technique is circular dichroism (CD). The matrix element for the rotational strength of a particular transition $\psi_i \rightarrow \psi_j$ is (see Eq. 5)

$$R_{i \rightarrow j} = \mathrm{Im}\{\langle i|\mathbf{M}|j\rangle \cdot \langle j|\mathbf{E}|i\rangle\} \tag{6}$$

where **M** and **E** are the magnetic and electric dipole moment operators, respectively. By contrast, absorption spectra depend on the magnitude of the electric dipole transition moment alone. Since the d orbital wavefunctions are even, the electric dipole term is forbidden, though some of the forbiddenness is removed by asymmetry in the molecular geometry, and this leads to the observed intensities of the electronic origins in absorption spectra. Mixing an odd vibration into the ground or excited state wavefunction (odd relative to the closest structure with a center of symmetry) creates an allowed transition (a vibronic satellite), which, depending on the efficiency of the mixing, may rival the electronic origin in intensity in the absorption spectrum. Broad band intensities in absorption spectra are almost entirely derived from this mechanism. The magnetic dipole transition between d-orbital wavefunctions is allowed, however, and this should in principle weight the pure electronic lines more heavily than the vibronic lines in the CD spectrum, relative to the absorption spectrum.

CD spectra of Cr(III) complexes in the spin-forbidden region do seem to show the expected simplicity when room temperature solution samples are used [51, 52], and the peaks have been associated with electronic origins. However,

Geiser and Güdel's 8 K spectrum of a crystalline sample of $[Cr(en)_3]Cl_3$ (en = ethylenediamine) exhibits fully as much vibronic structure as an absorption spectrum (Fig. 5) taken under the same conditions, and is displayed in Fig. 6 [37]. Though the relative line intensities are considerably different in the two spectra, there is no obvious advantage for the electronic transitions in the CD spectrum. It is possible that there is just as much vibronic intensity in room temperature solution spectra as in Geiser and Güdel's 8 K spectrum, but the much greater line widths obscure the structure. This would imply that assignments of peaks to pure electronic transitions in room temperature spectra may be mistaken. More solid state CD spectra and some 77 K solution spectra might help to clarify this point.

Another possible means of distinguishing electronic from vibronic lines is two-photon spectroscopy. The relevant matrix element for a single beam two-photon transition is

$$\langle i|[\mathbf{E}\times\mathbf{E}]|j\rangle \tag{7}$$

where the symmetric direct product of the electric dipole operator is indicated [53, 54]. In O_h symmetry this product spans A_{1g}, E_g, and T_{2g}. For d^3 complexes this means that a spin-forbidden transition can borrow intensity through spin orbit coupling from the $^4A_{2g} \rightarrow {}^4T_{1g}$ transition, but not from the $^4A_{2g} \rightarrow {}^4T_{2g}$. For example, the intensity of the $^4A_{2g} \rightarrow {}^2E_g$ transition depends on

$$\frac{\langle {}^4A_{2g}|[\mathbf{E}\times\mathbf{E}]|{}^4T_{1g}\rangle\langle {}^4T_{1g}|H_{SO}(T_{1g})|{}^2E_g\rangle}{E({}^4T_{1g}) - E({}^2E_g)} \tag{8}$$

where H_{SO} is the spin-orbit coupling operator, the spatial part of which has T_{1g} symmetry. Both terms in the numerator are nonvanishing, so the transition is allowed by a mechanism that does not depend on vibronic mixing. For lower symmetry complexes, intensity may also be borrowed from the $^4A_{2g} \rightarrow {}^4T_{2g}$ transition.

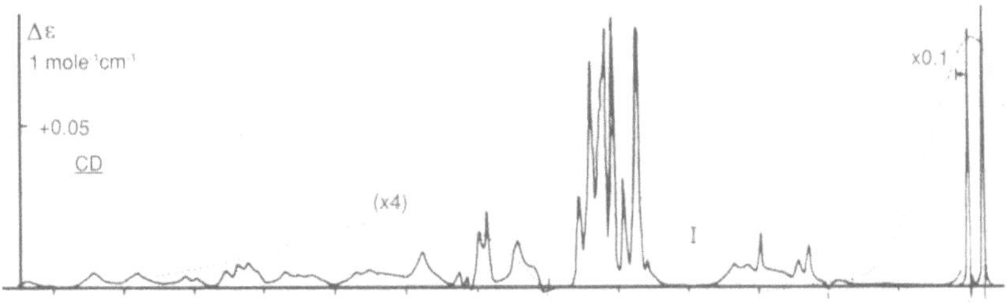

Fig. 6. Geiser and Güdel's axial circular dichroism spectrum of $[\Lambda\text{-}Cr(en)_3][\Delta\text{-}Ir(en)_3]Cl_6 \cdot KCl \cdot 6H_2O$ at 8 K [37]. The wavenumber scale is the same as in Fig. 5. The *dotted line* represents the room temperature solution spectrum of $[\Lambda\text{-}Cr(en)_3]^{3+}$. The *peak in the center* with a slightly negative Δε was assigned as an electronic origin

Again, since the d orbitals have even parity, even if the molecule does not have an inversion center there is an approximate selection rule in which transitions that would be g → g (or u → u) in a parent group with inversion symmetry are allowed. The odd parity vibrations that dominate the single photon spectrum are forbidden, while the even parity vibrations are allowed, but have no advantage over the pure electronic transitions. Experimental two-photon spectra of the sharp-line transitions of Mn^{4+} in a Cs_2GeF_6 host confirm both the simplicity of the spectrum and the relative prominence of the 0–0 lines [55].

If difficulties in reliably assigning electronic lines can be overcome, sharp-line spectroscopic techniques offer great promise in finding geometric information, especially in noncrystalline environments, and in determining geometry-independent AOM parameters that can be correlated with factors related to chemical bonding.

10 References

1. Schmidtke H-H (1964) Z Naturforsch A 19: 1502
2. Schäffer CE (1968) Struct Bonding 5: 68
3. Hoggard PE (1986) Coord Chem Rev 70: 85
4. Schmidtke H-H, Adamsky H, Schönherr T (1988) Bull Chem Soc Japan 61: 59
5. Lee K-W, Hoggard PE (1991) Inorg Chem 30: 264
6. Choi J-H, Hoggard PE (1992) Polyhedron 11: 2399
7. Hoggard PE, Lee K-W (1988) Inorg Chem 27: 2335
8. Clifford AA (1962) Multivariate error analysis, Wiley-Halstad, New York
9. Gerloch M, Slade RC (1973) Ligand-field parameters, Cambridge University Press
10. Hoggard PE (1988) Inorg Chem 27: 3476
11. Lancaster P, Tismenetsky M (1985) The theory of matrices, Academic, Orlando
12. Hoggard PE (1981) Z Naturforsch A 36: 1276
13. Decurtins S, Güdel HU, Neuenschwander K (1977) Inorg Chem 16: 796
14. Schläfer HL, Wagener H, Wasgestian HF, Herzog G, Ludi A (1971) Ber Bunsenges Phys Chem 75: 878
15. Flint CD, Palacio DJD (1977) J Chem Soc Faraday II 73: 649
16. Jagner S, Ljungstrom E, Vannerberg N-G (1974) Acta Chem Scand A 28: 623
17. Figgis BN, Reynolds PA, Williams GA (1981) Acta Cryst B 37: 504
18. Mukherjee RK, Bera SC, Bose A (1970) J Chem Phys 53: 1287
19. Mukherjee RK, Bera SC, Bose A (1972) J Chem Phys 56: 3720
20. Lee K-W, Hoggard PE (1990) Inorg Chem 29: 850
21. Tsirelson VG, Antipin MY, Gerr RG, Ozerov RP, Struchkov YT (1985) Phys Status Solidi 87: 425
22. Hauser A, Mäder M, Robinson WT, Murugesan R, Ferguson J (1987) Inorg Chem 26: 1331
23. Ceulemans A, Dendooven M, Vanquickenborne LG (1985) Inorg Chem 24: 1153
24. Ceulemans A, Dendooven M, Vanquickenborne LG (1985) Inorg Chem 24: 1158
25. Lee K-W, Hoggard PE (1989) Chem Phys 135: 219
26. Atanasov MA, Schönherr T, Schmidtke H-H (1987) Theor Chim Acta 71: 59
27. Schäffer CE, Yamatera H (1991) Inorg Chem 30: 2840
28. Lee K-W, Hoggard PE (1989) Polyhedron 8: 1557
29. Stewart JJP (1983) QCPE bulletin, Program 455
30. Riccieri P, Zinato E, Damiani A (1987) Inorg Chem 26: 2667
31. Kirk AD, Ibrahim AM (1988) Inorg Chem 27: 4567

32. Ceulemans A, Bongaerts N, Vanquickenborne LG (1987) Inorg Chem 26: 1566
33. Hoggard PE (1991) Inorg Chem 30: 4664
34. Hoggard PE (1991) 9th International symposium on photochemistry and photophysics of coordination compounds, Fribourg, Switzerland, Abstracts P67
35. Condon EL (1937) Rev Mod Phys 9: 432
36. Ballhausen CJ (1962) Introduction to ligand field theory, McGraw-Hill, New York
37. Geiser U, Güdel HU (1981) Inorg Chem 20: 3013
38. Kaizaki S, Hidaka J, Shimura Y (1973) Inorg Chem 12: 142
39. Hilmes GL, Brittain HG, Richardson FS (1977) Inorg Chem 16: 528
40. Whuler A, Brouty C, Spinat P, Herpin P (1977) Acta Cryst B 33: 2877
41. Lee K-W, Hoggard PE (1991) Transition Metal Chem 16: 377
42. Lee K-W (1989) Electronic spectroscopy and ligand field analysis of polypyridine and carboxylatochromium(III) complexes. Thesis, North Dakota State University
43. Schäffer CE, Jørgensen CK (1965) Kgl Danske Vidensk Selsk 34: No. 13
44. Purcell KF, Kotz JC (1977) Inorganic chemistry, Saunders, Philadelphia
45. Strand D, Linder R, Schmidtke H-H (1990) Mol Phys 71: 1075
46. Dingle RM, McCarthy PJ, Ballhausen CJ (1969) J Chem Phys 50: 1957
47. McClure DS (1962) J Chem Phys 36: 2757
48. Flint CD, Matthews AP (1973) J Chem Soc Faraday II 69: 419
49. Lee K-W, Hoggard PE (1988) Inorg Chem 27: 907
50. Riesen H (1988) Inorg Chem 27: 4677
51. Kaizaki S, Ito M, Nishimura N, Matsushita Y (1985) Inorg Chem 24: 2080
52. Kaizaki S, Mizu-uchi H (1986) Inorg Chem 25: 2732
53. Campochiaro C, McClure DS, Rabinowitz P, Dougal S (1989) Chem Phys Lett 157: 78
54. McClain WM (1974) Acc Chem Res 7: 129
55. Chien R-L, Berg JM, McClure DS, Rabinowitz P, Perry BN (1986) J Chem Phys 84: 4168
56. Boeyens JCA, Forbes AGS, Hancock RD, Wieghardt K (1985) Inorg Chem 24: 2926

Competition Between Ligand Centered and Charge Transfer Lowest Excited States in bis Cyclometalated Rh^{3+} and Ir^{3+} Complexes

Mirco G. Colombo, Andreas Hauser and Hans U. Güdel

Institut für Anorganische, Analytische und Physikalische Chemie, Universität Bern, Freiestrasse 3, CH-3000 Bern 9, Switzerland

Table of Contents

In this article the results of a detailed optical spectroscopic investigation of a series of related bis-cyclometalated Rh^{3+} and Ir^{3+} complexes of the general formula $[M(C^\frown N)_2N^\frown N]^+$, M = Rh^{3+}, Ir^{3+} ($HC^\frown N$ = 2-phenylpyridine or 2-(2-thienyl) pyridine; $N^\frown N$ = 2,2′-bipyridine or ethylenediamine) are summarized. The nature of the lowest excited states of the compounds is

Topics in Current Chemistry, Vol. 171

discussed on the basis of their absorption, luminescence and luminescence line narrowing spectra in solutions and glasses at temperatures from 10 K to 300 K, whereas for the characterization of higher excited states single crystal absorption spectra and excitation spectra of polycrystalline samples are used. In the Rh^{3+} complexes the lowest excited states correspond to a $^3\pi$–π^* transition localized on the cyclometalating ligands, whereas in the Ir^{3+} complexes, depending on the environment, either a $^3\pi$–π^* or a metal to ligand charge transfer (^{3}MLCT) excited state is lowest in energy. The pronounced solvatochromic and rigidochromic effects of the Ir^{3+} compounds are responsible for the reversal of the order of the lowest excited states. Mixing between the π–π^* and MLCT excited states is reflected in the oscillator strengths, luminescence lifetimes, vibrational structure and zero field splittings. The phenomenon of dual luminescence is attributed to a large inhomogeneous distribution of sites in solution and glasses.

1 Introduction

In the past a lot of preparative and spectroscopic work has been done on Ru^{2+} polypyridine complexes (e.g. $[Ru(bpy)_3]^{2+}$, bpy = 2,2′-bipyridine), because they have been and still are considered model systems for photo-sensitizing and solar energy conversion [1–4]. In search of new compounds with similar properties the central Ru^{2+} has been replaced by other d^6 ions, especially Rh^{3+} [5], and the ligand system has been modified, focussing on the substitution of one or more of the polypyridine ligands by cyclometalating ligands. The higher ligand field strength of the coordinating C^- as compared to N and the different donor/acceptor properties of these ligands can be used to tune the electronic properties of the complexes. The growing interest in cyclometalated compounds has resulted in a number of review articles treating the topic of cyclometalation from different points of view [6–14].

Among the ions with d^6 electron configuration Ir^{3+} and Rh^{3+} show a unique tendency for spontaneous cyclometalation. With bidentate aromatic ligands bis cyclometalated complexes are preferentially formed [15–26]. In contrast, only a few mono cyclometalated Rh^{3+} [27, 28] and Ir^{3+} [29–33] complexes are known, and for a long time tris-cyclometalated complexes have been restricted to Ir^{3+} [19, 34, 35]. Only very recently both meridional [36] and facial [36, 37] Rh^{3+} complexes have been synthesized.

The photocatalytic properties of a coordination compound are mainly determined by its lowest excited states. For complexes of the type under consideration the excited states can be classified as metal centered (MC, *d*–*d* transitions), ligand centered (LC, π–π^* transitions) and metal to ligand charge transfer states (d–π^*, MLCT transitions). In free ligands n–π^* as well as π–π^* lowest excited states are observed, but in the complexes treated here the formerly nonbonding orbitals are involved in the metal-ligand σ-bond, and we only consider π–π^* LC excitations. Low energy MLCT transitions are considered to be essential for the photocatalytic activity of the complexes. The assignment of the excited states as MC, MLCT and LC can often be done on the basis of intensities and band shapes of solution absorption spectra, and the nature of the lowest excited state can be inferred from solution luminescence spectra and excited state lifetimes. The active ligand in mixed ligand complexes can often be identified by comparing the luminescence band energies and electrochemical measurements. When excited states of a different nature lie close to each other, the information content of the generally broad solution spectra is too low to allow clear-cut assignments, and in particular such phenomena as solvent dependent band shapes and dual emission cannot be satisfactorily explained. Using high resolution spectroscopic methods, preferably in the solid state at cryogenic temperatures, it is possible to get a more detailed picture of the nature of the lowest excited states in such complicated situations.

In this article we summarize the results of our spectroscopic work on the bis cyclometalated complexes shown in Fig. 1. The ligands and complexes will be abbreviated as shown in Table 1. The complexes are all in a configuration with the coordinating carbons cis to each other and nitrogens of the cyclometalating ligands in trans position. In this configuration the complexes have a twofold symmetry axis, so the two cyclometalating ligands cannot be distinguished from NMR spectra in solution. For $[Rh(thpy)_2bpy]Cl \cdot 2^1/_8\ H_2O$ [25] and $[Rh(ppy)_2bpy]PF_6$ [38] the crystal structures have been determined. The powder X-ray diffraction data show that the Ir^{3+} complexes are isostructural with their corresponding Rh^{3+} complexes.

In our work we tried to combine solid state optical spectroscopy with the method of chemical variation, that is a systematic modification of the complexes (substitution of ligands and metals) as well as changes in the surrounding media (solvents, glasses and crystalline hosts). We are thus able to identify the nature of the excited states, and by studying the above series of related compounds we can discern the principles governing their energies relative to each other.

We begin by looking at the photophysical properties of our complexes in solution and glasses, discussing solvatochromism, rigidochromism as well as luminescence line narrowing. The characterization of the lowest excited states will be done on the basis of these investigations in disordered media. In the third section also higher excited states are examined by absorption spectroscopy of

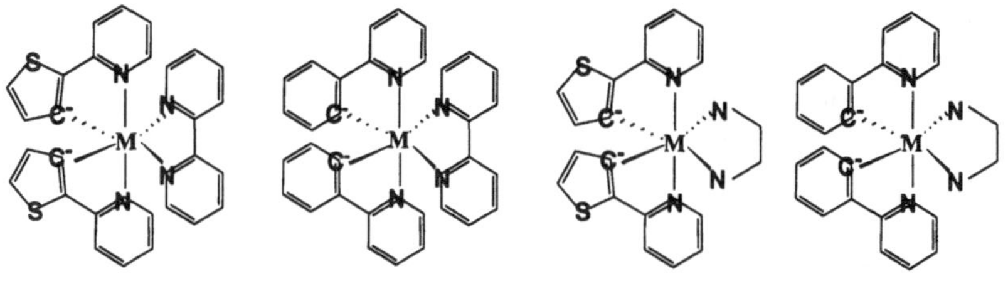

$MT2B^+$ $MP2B^+$ $MT2E^+$ $MP2E^+$

$M = Rh^{3+}, Ir^{3+}$

Fig. 1. Schematic structures of the bis-cyclometalated complexes treated in this article. The abbreviations refer to Table 1

Table 1. Formulas and abbreviations

bpy	2,2′-bipyridine
en	ethylendiamine
ppy^-	2-phenyl pyridine anion
$thpy^-$	2-(2-thienyl) pyridine anion
$RhT2B^+$	$(Rh(thpy)_2bpy)^+$
$RhT2E^+$	$(Rh(thpy)_2en)^+$
$RhP2B^+$	$(Rh(ppy)_2bpy)^+$
$RhP2E^+$	$(Rh(ppy)_2en)^+$
$IrT2B^+$	$(Ir(thpy)_2bpy)^+$
$IrT2E^+$	$(Ir(thpy)_2en)^+$
$IrP2B^+$	$(Ir(ppy)_2bpy)^+$
$IrP2E^+$	$(Ir(ppy)_2en)^+$

neat and doped single crystals as well as excitation and luminescence experiments of polycrystalline samples. In the fourth section the effect of mixing between the $^3\pi$–π^* and MLCT excited states is discussed in terms of luminescence lifetimes, polarization of transition moments, vibrational structure and zero field splittings, whereas the problem of dual luminescence from different excited states is addressed in Sect. 5.

2 Solutions and Glasses

2.1 Solution Absorption

The solution absorption spectra of all our eight complexes, displayed in Fig. 2, show roughly the same features: spin allowed ^{1}MLCT bands between 22 000 cm^{-1} and 28 000 cm^{-1}, and intense ^{1}LC bands above 30 000 cm^{-1}. The spectra of the complexes with either bpy and or en as a third ligand can only be distinguished in two points. The ^{1}LC band of the bpy samples (marked by an asterisk in Fig. 2) is missing in the en samples, and the ^{1}MLCT bands of the latter are red shifted. A red shift of the ^{1}MLCT bands is also observed when Rh^{3+} is replaced by Ir^{3+}. Additionally, in the Ir^{3+} complexes these bands are smeared out to the low energy side of the spectrum. This can be explained by low lying ^{3}MLCT states which get their intensity through efficient mixing with higher lying singlet states due to the larger spin-orbit coupling constant of Ir^{3+}.

The absorption bands can easily be classified as MLCT or LC based on their extinction coefficients and band energies. But assigning them to specific ligands in mixed ligand complexes can be quite difficult or impossible on the basis of solution spectra. Ethylenediamine does not have an extended aromatic system

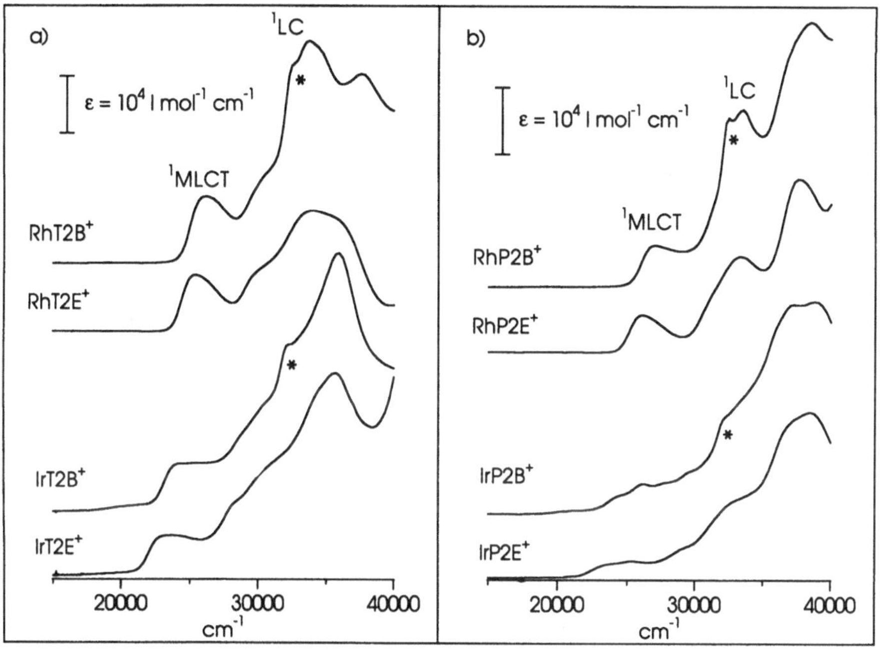

Fig. 2. Room temperature absorption spectra of a CH_2Cl_2 solution of **a** $[Rh(thpy)_2bpy]^+$, $[Rh(thpy)_2en]^+$, $[Ir(thpy)_2bpy]^+$ and $[Ir(thpy)_2en]^+$ and **b** $[Rh(ppy)_2bpy]^+$, $[Rh(ppy)_2en]^+$, $[Ir(ppy)_2bpy]^+$ and $[Ir(ppy)_2en]^+$. The *asterisks* mark the band assigned to a $^1\pi$–π^* transition on the bpy ligand

which could give rise to low energetic π–π^* transitions or stabilize by delocalization the formal extra charge provided in a MLCT transition. So in the complexes with en as chelating ligand the lowest energy excitations must involve the cyclometalating ligands. In the complexes with bpy as chelating ligand the electronic properties of the two ligand types are too similar to allow a reliable assignment of the broad absorption bands to a single ligand type. In particular, no further information can be obtained from the solution absorption spectra regarding weak ^{3}LC and ^{3}MLCT transitions expected below the intense ^{1}MLCT bands.

2.2 Solution Luminescence

For molecules in condensed media the luminescence generally originates from the lowest excited state or from higher excited states which are in thermal equilibrium with the lowest excited state. This makes the luminescence spectra a sensitive probe for the lowest excited states.

All four complexes with $thpy^-$ ligands show luminescence at room temperature in solution, as can be seen in Fig. 3a. The spectra of $[Rh(thpy)_2bpy]^+$,

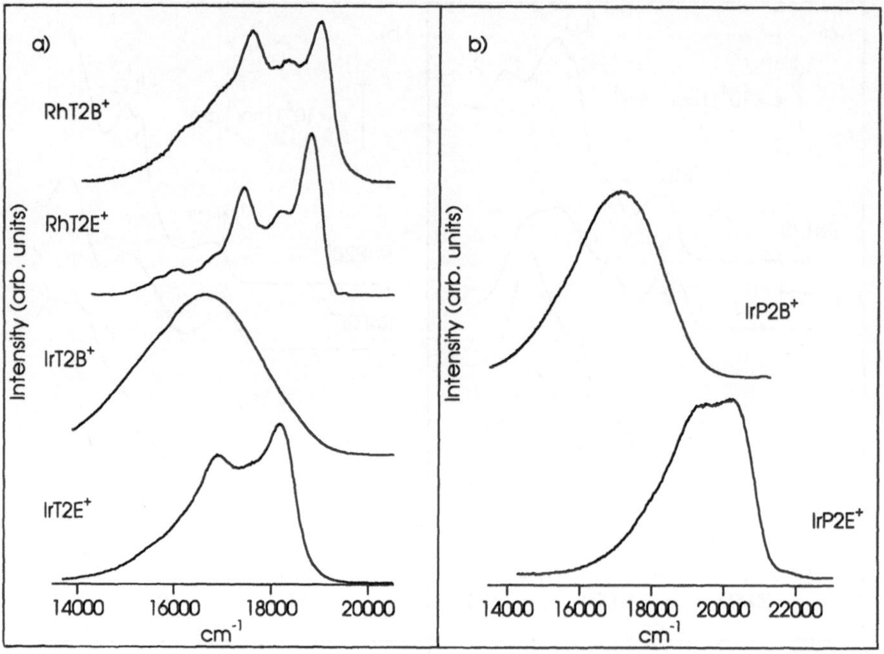

Fig. 3. Room temperature luminescence spectra of a CH_2Cl_2 solution of **a** $[Rh(thpy)_2bpy]^+$, $[Rh(thpy)_2en]^+$, $[Ir(thpy)_2bpy]^+$ and $[Ir(thpy)_2en]^+$ and **b** $[Ir(ppy)_2bpy]^+$ and $[Ir(ppy)_2en]^+$. The luminescence of $[Rh(ppy)_2bpy]^+$ and $[Rh(ppy)_2en]^+$ is not measurable under these conditions

$[Rh(thpy)_2en]^+$ and $[Ir(thpy)_2en]^+$ in CH_2Cl_2 are very similar: structured bands with intensity maxima between 18 200 cm^{-1} and 19 000 cm^{-1}. They have been assigned to ligand centered $^3\pi$–π^* transitions on the $thpy^-$ ligands [39–43], an assignment which will be confirmed in the following sections. Surprisingly, the luminescence of $[Ir(thpy)_2bpy]^+$ in CH_2Cl_2 has a completely different appearance: a broad unstructured band with its intensity maximum centred at 16 700 cm^{-1}. Its shape is characteristic for a ^{3}MLCT transition [26].

Of the complexes with ppy^- as cyclometalating ligand only the Ir^{3+} compounds have a measurable luminescence in solution at room temperature, as shown in Fig. 3b. The luminescence spectrum of $[Ir(ppy)_2bpy]^+$ in CH_2Cl_2 is almost identical to the one of $[Ir(thpy)_2bpy]^+$ and therefore it can also be assigned to a ^{3}MLCT transition, but its maximum is slightly blue shifted to 17 090 cm^{-1}. The similarity of the two spectra suggests that they originate from the same type of excited state. With bpy as the only common ligand an Ir → bpy MLCT assignment is indicated. We will show in Sect. 3 that this assignment is firmly established by polarized single crystal absorption spectroscopy. Furthermore this assignment is supported by cyclic voltammetry data of $[Ir(ppy)_2bpy]^+$ showing that bpy is more easily reduced than ppy^- [21] and consequently the lowest energy MLCT transition in $[Ir(ppy)_2bpy]^+$ should indeed involve the bpy ligand. In an excited state absorption study by Ichimura et al. [44] this result was confirmed by transient absorption bands which are characteristic for a $bpy^{-\cdot}$ radical. Despite these facts a lowest energy MLCT state involving the cyclometalating ligands has been postulated by Maestri et al. [39] for $[Rh(thpy)_2bpy]^+$ and $[Rh(ppy)_2bpy]^+$ on the basis of the solution absorption spectra.

The spectrum of $[Ir(ppy)_2en]^+$ in CH_2Cl_2 in Fig. 3b, consisting of a broad band with some structure superimposed, is intermediate between the typical structured bands of ^{3}LC transitions and the broad, unstructured bands of ^{3}MLCT transitions. With 20 240 cm^{-1} its emission maximum is at markedly higher energy than the emission maxima of the other complexes CH_2Cl_2.

^{3}MLCT bands typically show a pronounced solvatochromic effect which is usually ascribed to the large dipole moment of the MLCT state, whereas only a small effect is expected for ^{3}LC transitions. In fact the structured luminescence spectra of $[Rh(thpy)_2bpy]^+$, $[Rh(thpy)_2en]^+$ and $[Ir(thpy)_2en]^+$ are almost solvent independent, supporting their assignment as ^{3}LC transitions. As expected the spectral positions of the MLCT bands of $[Ir(thpy)_2bpy]^+$ and $[Ir(ppy)_2bpy]^+$ are strongly solvent dependent [26, 45]. The luminescence band of $[Ir(ppy)_2en]^+$ too, is solvent dependent, but the observed shifts are somewhat smaller than for the other two systems [46]. Only a tentative assignment as $Ir \rightarrow ppy^-$ MLCT transition is possible at this stage.

Using theories which relate the solvent induced band shifts to the dipole moments of the complexes [47–49] Wilde and Watts [50] estimated the transition dipole moment of $[Ir(ppy)_2bpy]^+$ and other bis-cyclometalated Ir^{3+} complexes. The results of their calculation will be discussed in Sect. 5.

2.3 Low Temperature Luminescence in Glasses

The low temperature luminescence spectra of the title compounds in poly(methyl methacrylate) (PMMA) are shown in Fig. 4. In addition to the solvatochromic effect the MLCT luminescence bands show a pronounced rigidochromic effect. However, not only the emission energies, but also the band shapes change when the low viscosity solvents are replaced by a rigid glassy matrix.

Replacing the CH_2Cl_2 solution by PMMA the spectrum of $[Ir(thpy)_2bpy]^+$ for instance changes its appearance from a ^{3}MLCT emitter to a ^{3}LC band with its intensity maximum at 18 293 cm^{-1}, whereas the ^{3}LC bands of the other $thpy^-$ compounds only show minor blue shifts. In PMMA at 10 K the luminescence spectra of all the $thpy^-$ complexes have similar shapes and energies (see Fig. 4) and therefore under these conditions they all have a $^3\pi$–π^* ($thpy^-$) lowest excited state.

The complexes with ppy^- as cyclometalating ligand show similar changes. The luminescence of the Rh^{3+} complexes, which was quenched in solution at room temperature, can be recorded in PMMA at 10 K and shows structured

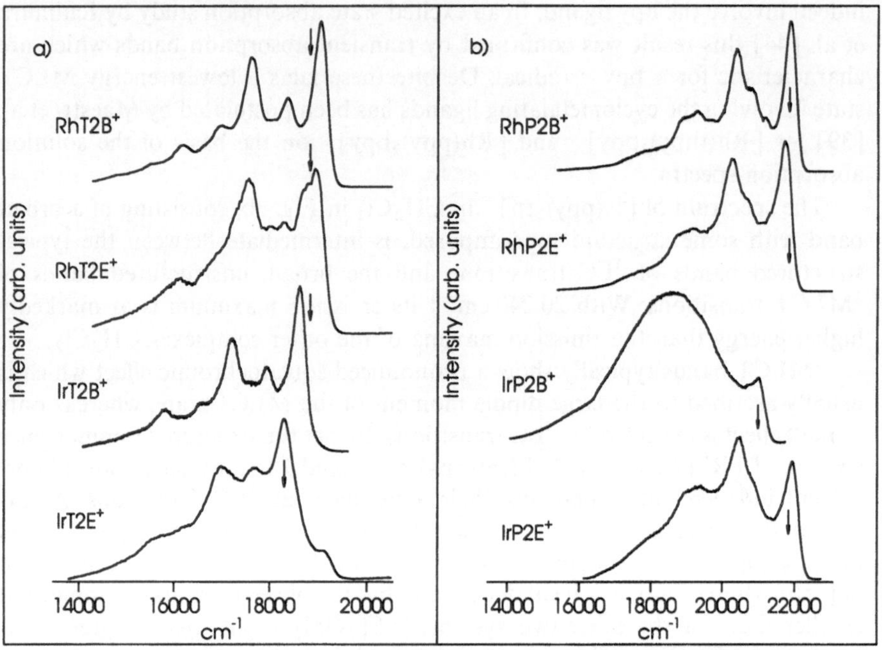

Fig. 4. Luminescence spectra at T = 10 K of **a** $[Rh(thpy)_2bpy]^+$, $[Rh(thpy)_2en]^+$, $[Ir(thpy)_2bpy]^+$ and $[Ir(thpy)_2en]^+$ and **b** $[Rh(ppy)_2bpy]^+$, $[Rh(ppy)_2en]^+$, $[Ir(ppy)_2bpy]^+$ and $[Ir(ppy)_2en]^+$ in poly (methyl methacrylate) (PMMA). The *arrows* indicate the position of the exciting laser lines used to generate the luminescence line narrowing spectra of Figs. 5 and 6

bands with their highest energy features around 21 800 cm^{-1}. Because of their identical shape and similar energy we can assign them to a $^3\pi$–π^* excitation on ppy^-, in accordance with Ref. [39].

The spectrum of $[Ir(ppy)_2en]^+$ in PMMA at 10 K looks similar to one of the analogous Rh^{3+} complex and can immediately be ascribed to a $^3\pi$–π^* transition on the ppy^-. It also has changed its behaviour from ^{3}MLCT type in CH_2Cl_2 at room temperature to ^{3}LC type in the glass at 10 K. But on closer inspection one suspects that there is still a broad band hidden under the structured luminescence. $[Ir(ppy)_2bpy]^+$, on the other hand, keeps its typical ^{3}MLCT luminescence band shape, but the band maximum shifts from 17 100 cm^{-1} in CH_2Cl_2 at room temperature up to 18 550 cm^{-1} in PMMA and, in addition, at low temperature a structured band shows up at 21 000 cm^{-1}. This structured band is significantly red shifted with respect to the ^{3}LC bands of the analogous Rh^{3+} complex and $[Ir(ppy)_2en]^+$, and so its nature cannot be determined at this stage.

In summary, all the Rh^{3+} complexes exclusively show ^{3}LC luminescence, whereas the Ir^{3+} compounds change their luminescence behaviour from dominantly ^{3}MLCT type in solutions at room temperature to ^{3}LC type in glasses at 10 K. This change is continuous and there are situations, depending on solvent and temperature, where both luminescences can be seen simultaneously. This can be rationalized by two close lying excited states, one ^{3}LC in character which is more or less independent to changes in the environment and one ^{3}MLCT in character which shifts its energy depending on the surrounding. In solutions the ^{3}MLCT state lies below the ^{3}LC state, whereas in rigid matrices this order is reversed at least for a subset of complexes. If the inhomogeneous width of the distribution of sites in the glass or in solution is larger than the mean energy separation between the two states, there will be complexes with a ^{3}MLCT and others with a ^{3}LC lowest excited state present in this medium. This aspect will be discussed in more detail in Sect. 5.

2.4 Luminescence Line Narrowing

Luminescence line narrowing (LLN) is a phenomenon which has been studied extensively in organic glassy matrices [51] as well as rare earth metal ions in ionic glasses [52]. There have only been a small number of LLN studies on coordination compounds. Riesen et al. [53] have recently reviewed the LLN and spectral hole burning work on transition metal complexes. The application of the LLN technique to this class of compounds can lead to an unambiguous identification on the ligand involved in the lowest excited state. Only a small subset of molecules of the broad inhomogeneous distribution of sites in a glass are excited by a narrow laser line tuned into the origin region. At low temperatures the resulting luminescence of these selectivity excited complexes is narrowed and the vibrational sidebands can be resolved. The vibrational sidebands are characteristic for each type of ligand, therefore they can be used as a fingerprint for the identification of the active ligand.

In Fig. 5 the luminescence line narrowing spectra of our title compounds in PMMA at $T = 6$ K are shown relative to the exciting laser line. Both the origin and the vibrational sidebands are accompanied by phonon wings. Their relative intensity depends on variables such as the medium, the temperature and the excitation wavelength. These effects have been discussed in Ref. [41]. Here we ignore the phonon sidebands and focus on the sharp line pattern. The spectra of the complexes with thpy$^-$ as cyclometalating ligand have almost identical patterns, see Fig. 5a. Replacement of bpy by en does not change the vibrational structure, and, with the exception of additional bands around 300 cm^{-1} in the Ir^{3+} spectra, the spectra of the Rh^{3+} and Ir^{3+} complexes are also identical. This is the most direct proof that the lowest energy excitation involves the thpy$^-$ ligand [40]. Furthermore it was shown, that the vibrational frequencies are independent of the counterion, the imbedding matrix and the position of the exciting laser line [40, 41].

The sideband patterns of the luminescence line narrowing spectra of $[Rh(ppy)_2bpy]^+$ and $[Rh(ppy)_2en]^+$ also coincide, see Fig. 5b, but are distinctively different from the ones of the thpy$^-$ containing samples, thus verifying that the lowest energy excitations involve the ppy$^-$ ligands [40, 41]. In

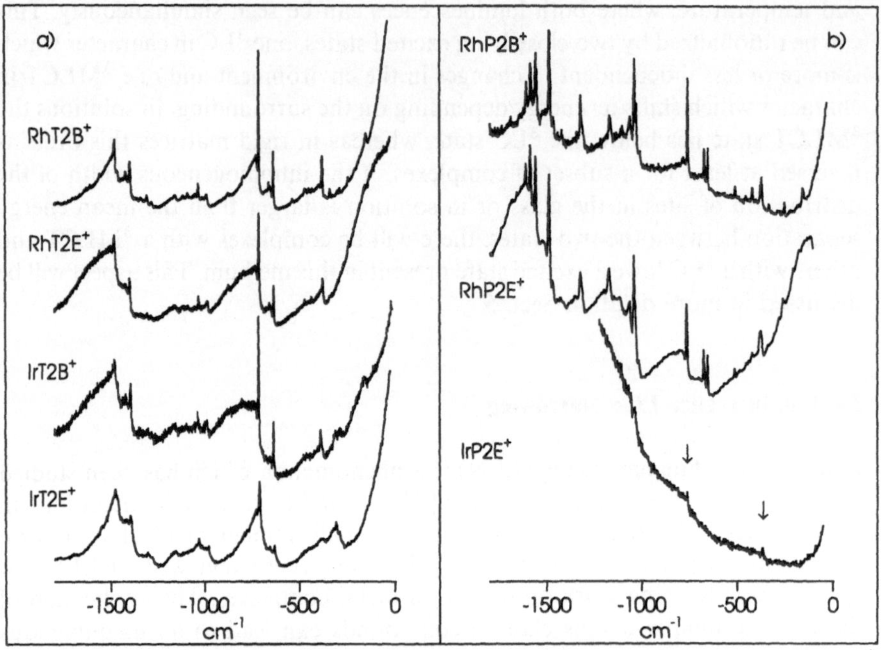

Fig. 5. Luminescence line narrowing spectra of **a** $[Rh(thpy)_2bpy]^+$, $[Rh(thpy)_2en]^+$, $[Ir(thpy)_2bpy]^+$ and $[Ir(thpy)_2en]^+$ and **b** $[Rh(ppy)_2bpy]^+$, $[Rh(ppy)_2en]^+$ and $[Ir(ppy)_2en]^+$ in PMMA at T = 6 K. The spectra are all displayed relative to the exciting laser line. They were excited at the arrow positions of Fig. 4. The *arrows* in the spectrum of $[Ir(ppy)_2en]^+$ mark weak vibrational sidebands

$[Ir(ppy)_2en]^+$, only partial line narrowing can be achieved, see arrows in Fig. 5b. The frequencies of the weak vibrational sidebands are very similar to the corresponding Rh^{3+} complexes, thus accounting for an excitation on the ppy^- ligands. No luminescence line narrowing could be observed for $[Ir(ppy)_2bpy]^+$.

The luminescence line narrowing technique can be applied to the ^{3}LC emission, where irradiation into the electronic origin is possible. No luminescence line narrowing could be observed in the complexes with ^{3}MLCT emissions. As a consequence the ^{3}MLCT part of the spectra is relatively more intense because the selectivity in the excitation process is greatly reduced. It should be noted, however, that Riesen et al. [54] achieved a narrowing of the ^{3}MLCT emission of $[Ru(bpy)_3]^{2+}$. The different effect of selective excitation on the Rh^{3+} and Ir^{3+} luminescence spectra is displayed in Fig. 6. The structured ^{3}LC band of $[Rh(ppy)_2en]^+$ is remarkably narrowed under selective excitation and sharp vibrational sidebands can be observed up to 3000 cm^{-1}. In $[Ir(ppy)_2en]^+$, only a minor part of the spectrum is narrowed leading to the weak vibrational structure in the low energy region. The main intensity belongs to a broad band similar in shape and energy to the ^{3}MLCT band of this complex in solution. In the luminescence line narrowing spectrum of $[Ir(ppy)_2bpy]^+$, the

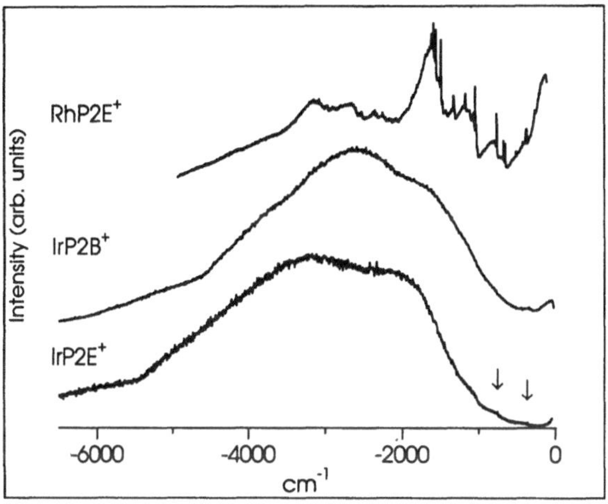

Fig. 6. Comparison of the selectively excited luminescence spectra of $[Rh(ppy)_2en]^+$, $[Ir(ppy)_2bpy]^+$ and $[Ir(ppy)_2en]^+$ in PMMA at T = 6 K

structured band overlaying the broad emission has disappeared. It also has been narrowed but the resulting sidebands are too weak to be detected on the strong underlying ^{3}MLCT band.

2.5 Summary

The LLN spectra clearly show that in our series of complexes the lowest ^{3}LC excited states involve the cyclometalating ligands. In the Rh^{3+} complexes this ^{3}LC exited state is always the lowest excited state, whereas in the analogous Ir^{3+} complexes a ^{3}MLCT excited state is very close in energy and can even cross the ^{3}LC state, depending upon the surrounding medium. In some cases both types of luminescence can be seen simultaneously. We ascribe this to the large inhomogeneous distribution of sites in the disordered matrices, giving rise to complexes with different lowest excited states. A full discussion of this so called phenomenon of dual luminescence will be given in Sect. 5.

3 Crystalline Samples

Crystals have two distinct advantages over amorphous samples: 1) the inhomogeneous broadening is reduced and 2) the complexes are oriented, and by using polarized light the principal components of the transition moment for the various absorption and emission lines can be determined. The sharp line absorption spectrum of $[Rh(ppy)_2bpy]PF_6$ is shown in Fig. 7. There are two close lying electronic origins C and D, with an identical vibrational sideband pattern built on them and a higher lying sharp origin N with a different pattern [38, 55]. The luminescence spectrum resulting after excitation in the UV region consists of very sharp lines too. Its origin is red shifted by 98 cm^{-1} with respect to the absorption origin C and coincides with a very weak absorption feature observed below the lines C and D [55]. Initially the luminescence was regarded as intrinsic [55, 56], but discrepancies between its vibrational structure and the Raman and luminescence line narrowing spectra of $[Rh(ppy)_2bpy]^+$ in a glass showed that it is a trap luminescence [43]. The weak absorption lines at lower energies are due to impurities, and the excitation energy migrates very rapidly through the crystal to these traps, so that only their luminescence can be observed. Extreme care is necessary when measuring and interpreting luminescence spectra of undiluted moleuclar coordination compounds of this type. The rate constant for excitation energy transfer between neighbouring complexes in the lattice is typically larger than $k = 10^9\ s^{-1}$. Both chemical impurities and crystal imperfections can act as excitation traps. There is also the possibility of deep traps which lead to a quenching of the luminescence.

The crystal structures of $[Rh(ppy)_2bpy]PF_6$ and $[Ir(ppy)_2bpy]PF_6$ are isomorphous as evidenced by their identical powder X-ray diffraction patterns. Therefore mixed crystals $[Ir_xRh_{1-x}(ppy)_2bpy]PF_6$, $0 \leq x \leq 1$ can be easily

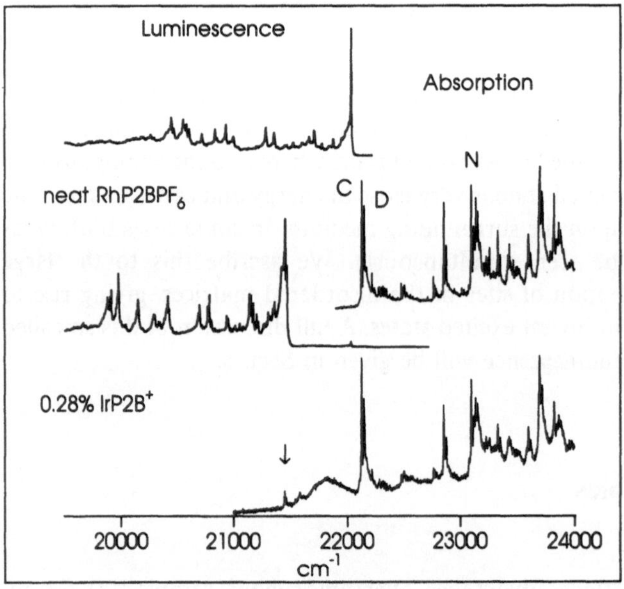

Fig. 7. Luminescence and absorption spectra of neat $[Rh(ppy)_2bpy]PF_6$ (*upper two spectra*) and of $[Ir(ppy)_2bpy]^+$ doped into $[Rh(ppy)_2bpy]PF_6$ (*lower two spectra*). The *arrow* indicates the sharp origin line of $[Ir(ppy)_2bpy]^+$. For the explanation of the lines *C*, *D* and *N* see text

prepared. The absorption spectrum of a dilute mixed crystal (x = 0.0028), displayed in the lowest trace of Fig. 7, shows additional sharp (arrow) and broad bands due to $[Ir(ppy)_2bpy]^+$ below the absorption origin of $[Rh(ppy)_2bpy]PF_6$. The luminescence of the doped crystal is shifted to lower energy with respect to the undoped host. The origin of the $[Ir(ppy)_2bpy]^+$ absorption coincides with the luminescence origin. The crystal was excited at 365 nm in this experiment, and we conclude that the excitation transfer from the Rh^+ host to the Ir^{3+} complexes is very efficient. Since the vibrational sideband pattern of the luminescence spectrum matches the one of the luminescence line narrowing spectrum of $[Rh(ppy)_2bpy]^+$ in PMMA at 6 K (see Fig. 5b), the lowest excited state of $[Ir(ppy)_2bpy]^+$ doped into $[Rh(ppy)_2bpy]PF_6$ is ascribed to a $^3\pi$–π^* transition on the ppy^- [45].

As shown in Fig. 7, not all the excitation is transferred to the $[Ir(ppy)_2bpy]^+$ trap in this mixed crystal. There is still a very small amount of the host luminescence left. The fraction of excitation energy which reaches the Ir^{3+} complexes depends on the temperature and the doping level. Transfer rates have been estimated by Zilian et al. [43].

Doped crystals can also be prepared if the neat dopant crystallises in a different structure than the host, but the charge and shape of the dopant should be appropriate and only low doping levels can be reached. $[Rh(thpy)_2bpy]^+$ and $[Ir(thpy)_2bpy]^+$ could in fact be doped into the crystal lattice of $[Rh(ppy)_2bpy]PF_6$. Their luminescence spectra with UV excitation are displayed in Fig. 8. With the exception of two features in the origin region the vibrational structure of the two spectra is identical and also coincides with the luminescence line narrowing spectra in PMMA, see Fig. 5a. This shows that also in the crystal the lowest excited state corresponds to a $^3\pi$–π^* transition on the $thpy^-$ for both complexes [26, 42, 57]. There are two distinguishing differences between the Rh^{3+} and Ir^{3+} spectra: The Stokes (pw_s) and anti-Stokes (pw_a) phonon wings accompanying the origin line and the vibrational sidebands are more intense in the Ir^{3+} spectrum than in the Rh^{3+} spectrum. Secondly there are two vibrations in the Ir^{3+} spectrum (arrows), which do not have a counterpart in the Rh^{3+} spectrum. An explanation of these observations will be given in the next section.

A big step forward in the understanding of the nature of the excited states of our compunds was provided by the polarised single crystal absorption spectra of neat $[Rh(ppy)_2bpy]PF_6$ in the work of Frei et al. [38]. From the crystal structure of $[Rh(ppy)_2bpy]PF_6$ and the dichronic ratios of the absorption bands the orientation of the transition moments of the three bands C, D and N (see Fig. 7) could be calculated with respect to the complex. It was concluded that C and D belong to $^3\pi$–π^* transitions localised on the ppy^- ligands whereas N corresponds to a $^3\pi$–π^* transition on the bpy. All three transitions are polarised within the plane of the respective ligand, which shows that they get most of their intensity through coupling to a higher ^{1}MLCT state. A pure π–π^* transition would be polarised out of plane [58, 59]. The splitting between the two origin lines C and D is a result of the crystal structure in which the two ppy^-

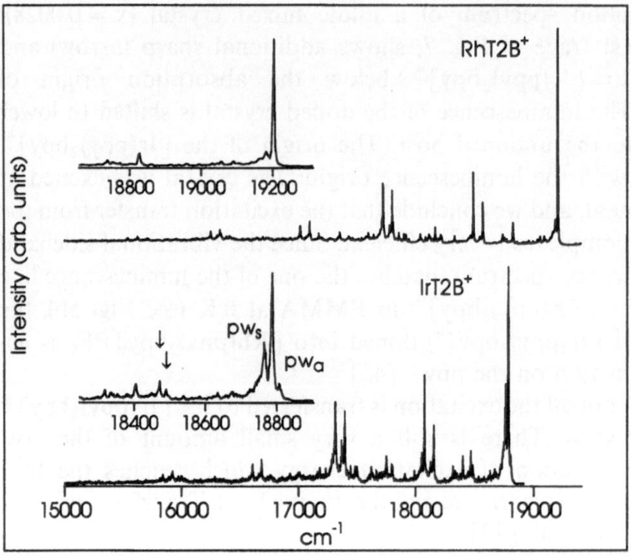

Fig. 8. Luminescence spectra at T = 10 K of $[Rh(thpy)_2bpy]^+$ (*upper trace*) and of $[Ir(thpy)_2bpy]^+$ (*lower trace*) doped into $[Rh(ppy)_2bpy]PF_6$. The *inserts* show the expanded origin region of the spectra. pw_s and pw_a label the Stokes or Antistokes phonon wings, respectively, which accompany the origin line and the vibrational sidebands. The *arrows* mark two vibrations in the Ir^{3+} spectrum without a counterpart in the spectrum of the corresponding Rh^{3+} complex

ligands are crystallographically inequivalent. It is a measure of how much the site geometry is distorted from C_2 symmetry [38].

The use of polarised single crystal spectra is not limited to neat systems. Also doped crystals can be investigated as shown in Fig. 9 for $[Ir(thpy)_2bpy]^+$ doped into $[Rh(ppy)_2bpy]PF_6$. There are two features due to $[Ir(thpy)_2bpy]^+$: two sharp origin lines, also labelled C and D with their corresponding vibrational sidebands, and a broad band centred at 21 700 cm^{-1} and labelled MLCT. Applying the same procedure as in ref. [38] we found that the sharp bands C and D belong to $^3\pi$–π^* transitions on the $thpy^-$ ligands and the broad band is polarized in the Ir → bpy direction [26]. Because of its shape and oscillator strength the broad band can be assigned to an Ir → bpy ^{3}MLCT transition.

Similar conclusions can be drawn from the absorption spectra of $[Ir(ppy)_2bpy]^+$ doped into $[Rh(ppy)_2bpy]PF_6$ (see Fig. 7). The sharp absorption bands around 21450 cm^{-1} correspond to $^3\pi$–π^* transitions on the inequivalent ppy^- ligands, and the broad band with its maximum at 21 820 cm^{-1} is a ^{3}MLCT transition to the bpy [45]. In contrast to $[Ir(thpy)_2bpy]^+$ the energy separation between the ^{3}LC and ^{3}MLCT states is much smaller in $[Ir(ppy)_2bpy]^+$.

From the absorption spectra the oscillator strengths f_{ij} of the various transitions listed in Table 2 can be derived. The oscillator strengths can be

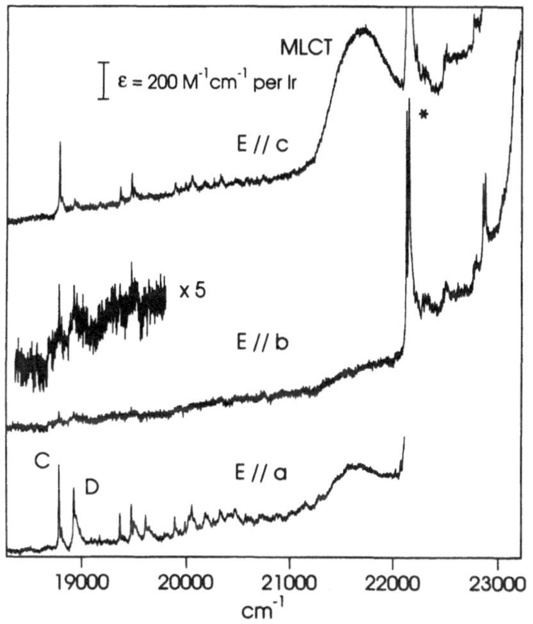

Fig. 9. Polarised single crystal absorption spectra at $T = 9\,K$ of $[Ir(thpy)_2bpy]^+$ doped into $[Rh(ppy)_2bpy]PF_6$. The spectra were recorded with the electric field vector parallel to the $\vec{c}$-, $\vec{b}$- and $\vec{a}$-axes of the crystal, respectively. The labels *C*, *D* and *MLCT* are explained in the text. The onset of the host absorption is marked by an *asterisk* (from Ref. [26])

correlated with the radiative rates k_{rad} according to [60, 61]:

$$f_{ij}(ed)\frac{1}{k_{rad}} = 1.5 \times 10^4 \frac{g_g}{g_e} \frac{(c/\nu)^2}{n\left(\frac{n^2+2}{3}\right)^2} \text{ [SI units]}, \tag{1}$$

where ν is the transition frequency, n the refractive index of the absorbing medium and c the vacuum velocity of light. g_g and g_e are the degeneracies of the ground and the excited state, respectively.

The calculated radiative lifetimes $\tau_{rad} = 1/k_{rad}$ of the lowest excited state for several compounds are included in Table 2. For the calculation a refractive index of 1.5 was assumed.

With the exception of $[Ir(ppy)_2en]^+$, where we have a very big uncertainty in the estimated absorption intensity, the correlation between the calculated radiative and observed lifetimes at low temperatures is excellent. This proves that at low temperatures the lifetime of the excited state is mainly governed by its radiative decay rate [26].

Often single crystals suitable for absorption spectroscopy cannot be grown, but the problem can be overcome by the use of excitation spectroscopy. With narrow band tuneable lasers highly resolved spectra of polycrystalline and opaque samples can be obtained.

The excitation spectra at 12 K of the bis-cyclometalated Rh^{3+} and Ir^{3+} complexes doped into $[Rh(ppy)_2bpy]PF_6$ and $[Rh(ppy)_2en]PF_6$ are displayed in Fig. 10. In all spectra the two origin lines C and D of the crystallographically

Table 2. Oscillator strengths f_{ij}, calculated (τ_{rad}) and experimental (τ_{exp}) lifetimes of the lowest excited states (C, D and MLCT) of the complexes and the protonated free ligands

Compound	Oscillator strength f_{ij}			Luminescence lifetime		
	^{3}LC(C⌒N$^-$): C	^{3}LC(C⌒N$^-$): D	^{3}MLCT(bpy)	τ_{rad} [s]	τ_{exp} [s]	T [K]
Hthpy					3.5×10^{-2} [b]	77
Hppy					> 0.1 [b)]	77
$Hbpy^+$					1.01 [b)]	77
$[Rh(thpy)_2bpy]^+$					5×10^{-4}	77
$[Rh(thpy)_2en]^+$					2.9×10^{-4}	77
$[Rh(ppy)_2bpy]^+$	1.69×10^{-5}	2.37×10^{-5}		1.5×10^{-4} [a]	1.7×10^{-4}	77
$[Rh(ppy)_2en]^+$	4.95×10^{-5}	1.7×10^{-4}		6.5×10^{-5}	6.7×10^{-5}	77
$[Ir(thpy)_2bpy]^+$	2.06×10^{-4}	1.53×10^{-4}	7.12×10^{-4}	2.1×10^{-5}	1.7×10^{-5}	10
$[Ir(thpy)_2en]^+$	5.44×10^{-4}	6.13×10^{-4}		8.3×10^{-6}	9.8×10^{-6}	40
$[Ir(ppy)_2bpy]^+$	3.82×10^{-4}	1.06×10^{-3}	1.68×10^{-3}	4.5×10^{-6} [a]	4.8×10^{-6}	77
$[Ir(ppy)_2en]^+$	6.9×10^{-3}	1.15×10^{-2}		5×10^{-7}	7.4×10^{-6}	45

[a] At the temperature where the experimental lifetime was measured, the C and D states are about equally populated. Therefore the average of their oscillator strengths was used for the calculation of the lifetime.
[b] From Ref. [39].

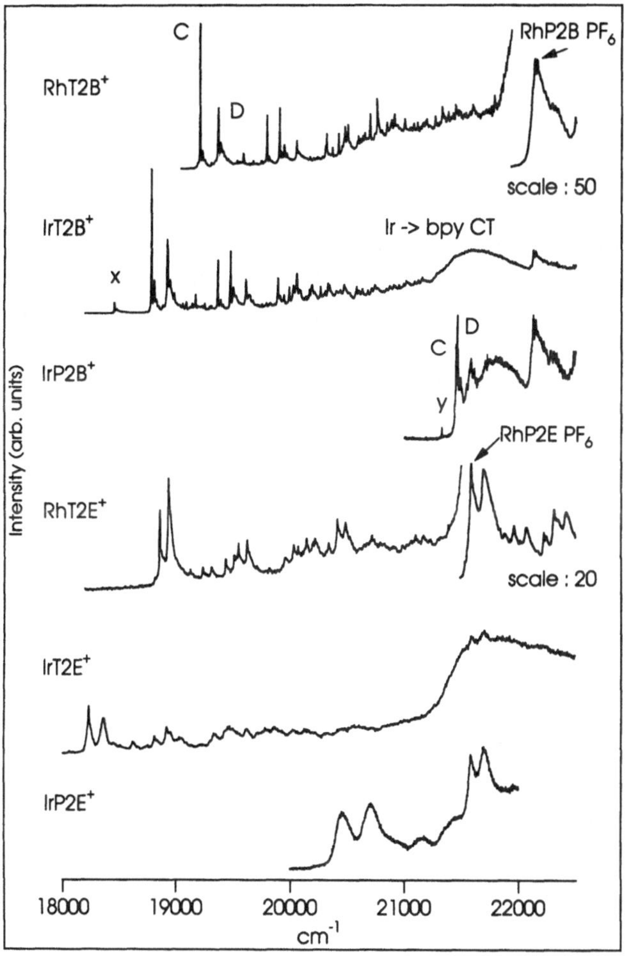

Fig. 10. Low temperature (T = 12 K) excitation spectra of $[Rh(thpy)_2bpy]^+$, $[Ir(thpy)_2bpy]^+$ and $[Ir(ppy)_2bpy]^+$ doped into $[Rh(ppy)_2bpy]PF_6$ (*upper three spectra*) and of $[Rh(thpy)_2en]^+$, $[Ir(thpy)_2en]^+$ and $[Ir(ppy)_2en]^+$ doped into $[Rh(ppy)_2en]PF_6$ (*lower three spectra*). *x* and *y* label traps

inequivalent cyclometalating ligands are at lowest energy. The frequencies and relative intensities of the vibrational sidebands of the four complexes with $thpy^-$ ligands are roughly the same, as expected for a $^3\pi$–π^* transition on the $thpy^-$. The spin forbidden Ir → bpy CT band at 21 700 cm^{-1} occurs at approximately the same energy in the spectra of $[Ir(thpy)_2bpy]^+$ and $[Ir(ppy)_2bpy]^+$, whereas the $^3\pi$–π^* (ppy^-) state is found 2500 m^{-1} above the $^3\pi$–π^* ($thpy^-$) state. In the spectrum of $[Ir(thpy)_2en]^+$ there appears to be a broad band at about 21 700 cm^{-1} as well. Although the situation is obscured by the absorption of the host at slightly higher energy, this band is tentatively ascribed to an Ir → $thpy^-$

^{3}MLCT transition. Generally the determination of higher excited state energies is very difficult. In doped crystals the bands are hidden under the host absorption and in neat crystals the extinction of charge transfer bands often exceeds the range of the apparatus.

Nevertheless, by choosing the proper dopant concentration and absorption pathlength also the broad absorption features of higher lying ^{3}MLCT transitions may be observed in neat and doped crystals as shown in Fig. 11 for $[Ir(thpy)_2bpy]^+$ doped into $[Rh(ppy)_2bpy]PF_6$ as well as neat $[Rh(ppy)_2en]PF_6$ and $[Rh(ppy)_2bpy]B\phi_4$ ($B\phi_4$ = tetraphenylborate). The sharp features in the absorption spectrum of Fig. 11a are due to the $[Rh(ppy)_2bpy]PF_6$ host. The corresponding $^3\pi$–π^* transitions of the dopant at lower energy are too weak in this polarization (E//b) to be observed on this scale. The shoulder at 23 200 cm^{-1} was ascribed to an Ir → $thpy^-$ ^{3}MLCT transition because the Ir → bpy ^{3}MLCT transition already has been observed at 21 700 cm^{-1} and the ^{3}MLCT absorptions of the neat host occur at distinctively higher energy. For $[Rh(ppy)_2en]PF_6$ (Fig. 11b) the assignment of the shoulder at 24 800 cm^{-1} to a Rh → ppy^- ^{3}MLCT transition is straightforward, whereas the assignment of the band at 25 000 cm^{-1} in the spectrum of $[Rh(ppy)_2bpy]B\phi_4$ to a Rh → bpy ^{3}MLCT transition was made in analogy to the Ir^{3+} complexes, where the Ir → bpy ^{3}MLCT transitions lie below the corresponding Ir → ppy^- or Ir → $thpy^-$ transitions.

The relevant excited state energies of all the complexies are compiled in Table 3 and visualised in Fig. 12.

We can observe some general trends: the $^3\pi$–π^* excited states of ppy^- lie about 3000 cm^{-1} above the ones of $thpy^-$, whereas $^3\pi$–π^* excited states of bpy are a further 1000 cm^{-1} higher in energy. The transitions of the complexes with en as the third ligand are at lower energy than the corresponding transitions of

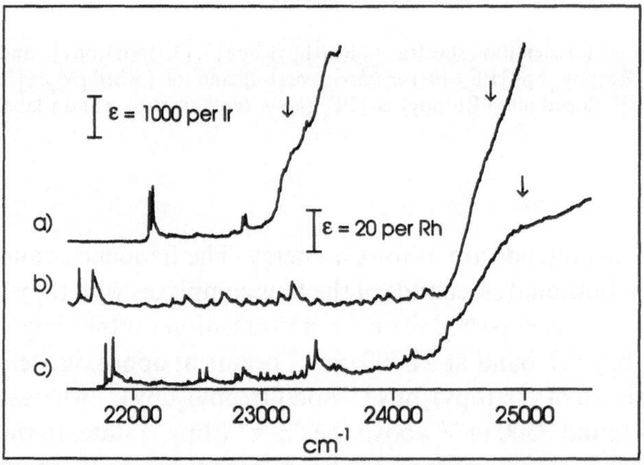

Fig. 11. Single crystal absorption spectra of $[Ir(thpy)_2bpy]^+$ doped into $[Rh(ppy)_2bpy]PF_6$ (a), neat $[Rh(ppy)_2en]PF_6$ (b) and neat $[Rh(ppy)_2bpy]B\phi_4$ (c) at T = 10 K

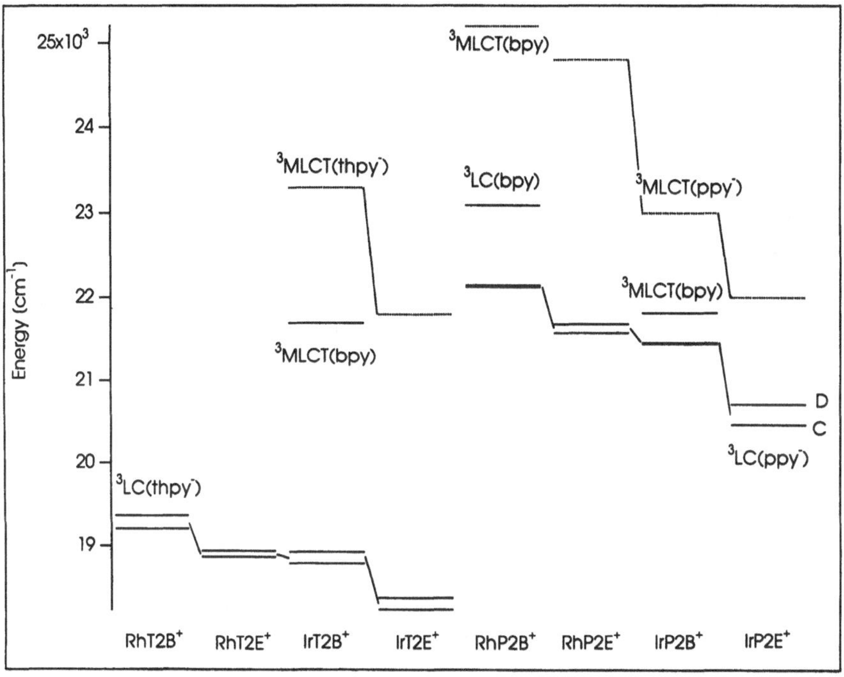

Fig. 12. Franck-Condon energies of the lowest excited states of the complexes as observed in the absorption/excitation spectra of crystalline samples. The *broken lines* characterize states of which the assignment or the exact energy have not been confirmed by other experiments yet

the complexes with bpy as the third ligand. States of the Ir^{3+} complexes lie below the corresponding states of the Rh^{3+} complexes. The $^3\pi$–π^* excited states of a ligand show comparatively small energy shifts on substitution of the metal or another ligand, whereas the ^{3}MLCT excited states change their energies drastically if Rh^{3+} is replaced by Ir^{3+} or the chelating ligand bpy by en. The fact that the lowest MLCT state of $[Ir(thpy)_2bpy]^+$ and $[Ir(ppy)_2bpy]^+$ involves the bpy and not the cyclometalating ligands points to a situation where the π^* orbitals of bpy lie below the π^* orbitals of the cyclometalating ligands as sketched in Fig. 13. This is supported by electrochemical measurements which put the reduction potential of bpy below the ones of ppy^- and $thpy^-$ [39]. If the σ-donor – π–acceptor bpy is replaced by the σ-donor en the $t_{2g}(\pi)$ orbitals move up in energy thus reducing the energy of the MLCT transitions on the cyclometalating ligands. This effect is less due to the different σ-donor strength of en acting mostly on the empty $e_g(\sigma)$ orbitals but rather to the loss of stabilization of the $t_{2g}(\pi)$ orbitals through π–backbonding. The labels in Fig. 13 represent the leading term in the linear combinations describing the molecular orbitals. Extended Hückel calculations on a $[Rh(ppy)_2bpy]^+$ fragment have shown, that in fact the LUMO is almost a pure bpy orbital, whereas for the HOMO a strong mixing of ppy^- character into the t_{2g} orbitals is found [63]. In this

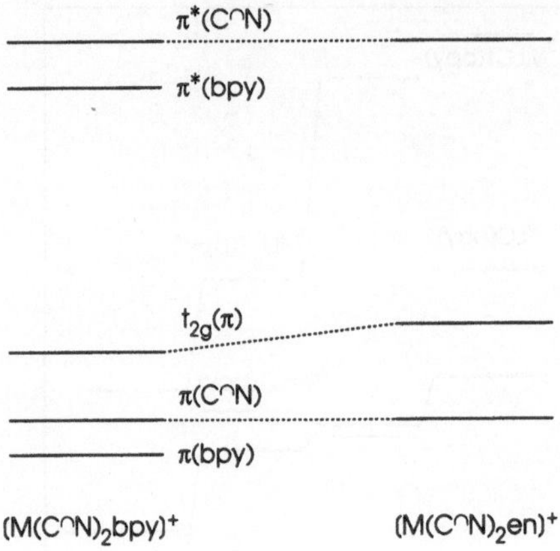

Fig. 13. Simplified molecular orbital scheme. For clarity the fully occupied lower lying and the empty higher lying orbitals have been omitted and the notation refers to O_h symmetry

picture the transitions classified as Rh → bpy MLCT have some formal ppy^- → bpy ligand-ligand charge transfer character.

This order of the molecular orbitals reproduces well the order of the singlet exited states. But it is the lower lying triplet states which determine the photophysical properties of the complexes. Therefore the energy separation between singlet and triplet states deriving from the excited π–π^* or t_{2g}–π^* electron configurations have to be taken into account. These separations are governed by electron repulsion integrals, which are significantly bigger for the π–π^* configuration, in which the unpaired electrons are spatially confined to a given ligand than for the t_{2g}–π^* configuration, in which the unpaired electrons are spatially separated. The singlet-triplet separations of the π–π^* states can be estimated from the shifts between the maxima of the respective absorption and luminescence bands. With the spectral values from Figs. 2 and 10 we get splittings of about 14 000 cm^{-1}, 13 000 cm^{-1} and 10 000 cm^{-1} for $thpy^-$ ppy^- and bpy, respectively. Singlet-triplet splittings of the MLCT excitations of approximately 4000 cm^{-1} are estimated from the solution and crystal absorption spectra of the complexes. With these values it becomes clear that the $^3\pi$–π^* excited states can lie below the corresponding ^{3}MLCT excited states even if the π orbital lies below the t_{2g} orbital in the one electron picture. The same argument holds for the order of the π^* orbitals of bpy and the cyclometalating ligands and the respective energies of their $^3\pi$–π^* excited states.

4 Mixing Between the LC and MLCT Excited States

The decrease in $^3\pi$–π^* luminescence lifetimes by several orders of magnitude in heavy metal complexes as compared to the lifetimes of the free ligand is usually attributed to mixing between the ^{3}LC and MLCT excited states [64]. This is borne out by the work of Frei et al. [38] which shows that the ^{3}LC transitions of the Rh^{3+} complexes get their intensity through mixing with the ^{1}MLCT states, thus leading to in-plane polarized transition moments. Comparing the complexes in Table 2, we notice that there is another one to two orders of magnitude decrease in the luminescene lifetimes between corresponding Rh^{3+} and Ir^{3+} complexes. This is clear evidence that the mixing through spin-orbit coupling is much stronger in the Ir^{3+} complexes than in the corresponding Rh^{3+} complexes. In fact the oscillator strengths of the ^{3}LC (ppy^-) and ^{3}MLCT (bpy) bands of $[Ir(ppy)_2bpy]^+$ are of the same order of magnitude. The increased mixing between the ^{3}LC and ^{1}MLCT excited states of the Ir^{3+} complexes is evident in other observables as well. In the highly resolved LLN as well as in the crystal luminescence and excitation spectra of the Ir^{3+} complexes (see Figs. 5 and 8) low energy vibrational sidebands around 300 cm^{-1} are found, which do not occur in the spectra of the corresponding Rh^{3+} complexes [26, 45]. This energy range is characteristic for metal-ligand vibrations [65], and calculations on a $Ru(bpy)^{2+}$ fragment showed, that these vibrations indeed carry a considerable amount of metal-ligand character [66]. Consequently the occurrence of metal-ligand vibrational sidebands in nominally ligand centred transitions is taken as evidence for their increased charge transfer character [26, 45]. In the well resolved spectra of the crystalline samples the phonon wings accompanying the sharp lines are more intense in the Ir^{3+} than in the Rh^{3+} spectra (see Fig. 8). The intensity ratio of the phonon wing and the sharp line is a measure for the coupling of the chromophore to the host [67]. MLCT states are known to depend more strongly on their environment than LC states. Therefore their larger phonon wings substantiate the increased charge transfer character of the nominal $^3\pi$–π^* states of the Ir^{3+} compounds.

The mixing of ^{1}MLCT character into ^{3}LC states is also known to increase the zero field splittings (ZFS) [64] of the latter. From combined Zeeman and hole burning [68] / luminescence line narrowing [42, 69] experiments on $[Rh(ppy)_2bpy]^+$ and $[Rh(thpy)_2bpy]^+$ a maximum ZFS $|D| + |E| \leq 0.4\ cm^-$ was estimated for the two complexes. This estimate was later substantiated by an ODMR investigation of $[Rh(thpy)_2bpy]^+$ doped into $[Rh(ppy)_2bpy]PF_6$ giving a total ZFS of 0.144 cm^{-1} [70]. This splitting is too small to be resolved in an optical absorption spectrum. The first direct observation of a ZFS in an optical spectrum of a bis-cyclometalated complex was made for $[Ir(ppy)_2bpy]^+$ doped into $[Rh(ppy)_2bpy]PF_6$. In the excitation and luminescence spectra a third origin line M is observed 10 cm^{-1} below the origin C (see Fig. 14) [45]. The temperature dependent intensity distribution between the three origin lines and the fact that the luminescence spectra are independent on selective excitation into the different origins show that we are dealing with an electronic splitting and not with complexes on different crystal sites. Therefore the line M was assigned to a spin sublevel of the origin C corresponding to a ZFS of about 10 cm^{-1}. A splitting of the origin line C is also observed in the low temperature excitation spectrum of $[Ir(thpy)_2bpy]^+$ doped into $[Rh(ppy)_2bpy]PF_6$ (see Fig. 15) [46]. With a value of about 5 cm^{-1} this ZFS is substantially smaller than the one of $[Ir(ppy)_2bpy]^+$.

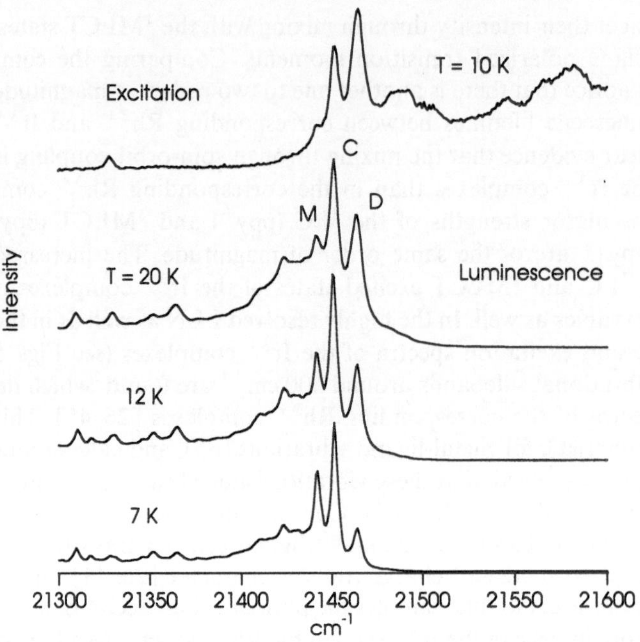

Fig. 14. Origin region of the low temperature excitation (luminescence monitored broad band below 20 200 cm^{-1}) and luminescence (excitation at 457.9 nm with an Ar laser) spectra of $[Ir(ppy)_2bpy]PF_6$ in $[Rh(ppy)_2bpy]PF_6$. *M*, *C* and *D* label electronic origins (from Ref. [45])

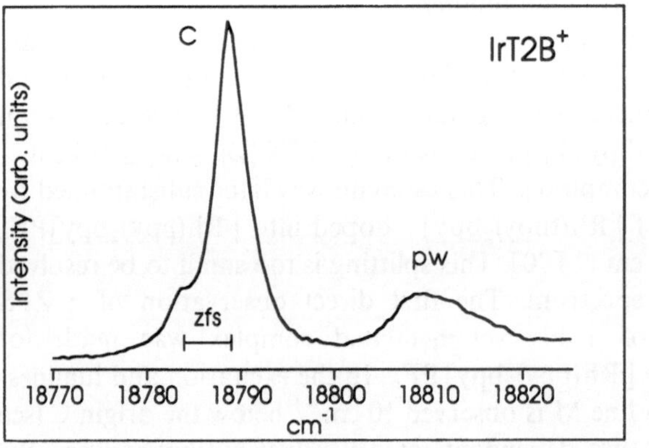

Fig. 15. Expanded origin region of the excitation spectrum of $[Ir(thpy)_2bpy]^+$ doped into $[Rh(ppy)_2bpy]PF_6$ at T = 6 K. *ZFS* indicates a zero field splitting of the origin line of approximately 5 cm^{-1}

We tried to rationalize the one to two orders of magnitude difference between the ZFS of the Rh^{3+} and the Ir^{3+} complexes. The contribution of the mixing between the ^3LC and ^1MLCT states to the ZFS is given by the second order perturbation expression [64]:

$$ZFS(^3LC) = |\langle ^1MLCT|H_{SO}|^3LC\rangle|^2 \times \left(\frac{1}{h\nu(^3MLCT) - h\nu(^3LC)} - \frac{1}{h\nu(^1MLCT) - h\nu(^3LC)}\right) \tag{2}$$

The spin-orbit coupling matrix element $\langle ^1MLCT|H_{SO}|^3LC\rangle$ can be calculated from the radiative rate constants and the energies of the states involved [64]:

$$\frac{k_{rad}(^3LC)}{k_{rad}(^1MLCT)} = \left(\frac{\nu(^3LC)}{\nu(^1MLCT)}\right)^3 \left(\frac{\langle ^1MLCT|H_{SO}|^3LC\rangle}{h\nu(^1MLCT) - h\nu(^3LC)}\right)^2 \tag{3}$$

The ^3LC states mix, in first order, with the MLCT states involving the same ligand. Therefore the MLCT state energies of the cyclometalating ligands have to be taken for the calculation. The matrix elements and ZFS shown in Table 4 were calculated using average ^1MLCT energies of 25 000 cm^{-1} for the Ir^{3+} and 26 000 cm^{-1} for the Rh^{3+} complexes estimated from solution absorption spectra, the radiative rate constants of Table 2, and the triplet state energies of Table 3.

The match between the calculated and observed total ZFS $|D| + |E|$ is reasonable, in particular the order of magnitude difference between the Rh^{3+} and Ir^{3+} complexes is well reproduced. The observed splittings can thus be explained by the mixing between the LC and the MLCT excited states.

The zero field splittings of similar Rh^{3+} and Ir^{3+} complexes fit well into this picture. For the ^3LC emitter $[Rh(bpy)_3](ClO_4)_3$ ODMR measurements gave a ZFS of 0.12 cm^{-1} [71] and for $[Ir(5,6\text{-}(CH_3)_2\text{-}phen)_2Cl_2]$ which also shows a ^3LC type luminescence a ZFS of 3 cm^{-1} was derived from Zeeman experiments [72].

The ZFS depends strongly on the spectral position of the ^3MLCT state. It has been noted, that the ZFS becomes large when the ^3MLCT and ^3LC state lie closer to each other than 5000 cm^{-1} [72]. This explains why the ZFS of $[Ir(thpy)_2bpy]^+$ and $[Ir(ppy)_2bpy]^+$ differ by a factor of two even though the matrix elements are of similar size.

5 Dual Emission

In the early 1970s, Demas and Crosby [73, 74] postulated that in transition metal complexes with unfilled *d* shells luminescence should only be observed from the lowest excited or thermally populated higher excited states. The condition for this thermalization is a fast radiationless relaxation between the excited states. It was then suggested by Watts et al. [75] that the relaxation between the states may be hindered if the excited states have a different orbital parentage and the energy difference between the states is small (< 300 cm^{-1}).

Table 3. Energies [cm^{-1}] of the lowest excited states as observed in absorption and excitation spectra of crystalline samples

Complex	Host	^{3}LC ($thpy^-$)		^{3}LC (ppy^-)		^{3}LC(bpy)	^{3}MLCT			Refs.
		C	D	C	D	N	$thpy^-$	ppy^-	bpy	
$[Rh(thpy)_2bpy]^+$	neat PF_6^- salt	18979[a]	19239[a]	–	–			–		[26]
	$[Rh(ppy)_2bpy]PF_6$	19205	19361	–	–			–		[42, 57]
	$[Ir(ppy)_2bpy]PF_6$	19209	19347	–	–			–		[26]
$[Rh(thpy)_2en]^+$	$[Rh(ppy)_2en]PF_6$	18865	18941	–	–	–		–	–	[46]
$[Ir(thpy)_2bpy]^+$	$[Rh(thpy)_2bpy]PF_6$	18557[a]	18665[a]	–	–			–		[26]
	$[Rh(ppy)_2bpy]PF_6$	18789	18928	–	–		23300[b]	–	21700	[26]
	$[Ir(ppy)_2bpy]PF_6$	18799	18915	–	–			–		[26, 46]
$[Ir(thpy)_2en]^+$	$[Rh(ppy)_2en]PF_6$	18234	18368	–	–	–	21700[b]	–		[46]
$[Rh(ppy)_2bpy]^+$	neat PF_6^- salt	–	–	22126	22148	23093	–		> 25400[c]	[38, 62]
	neat $B\phi_4^-$ salt	–	–	21793	21851	22911	–		25000[c]	[38, 62]
$[Rh(ppy)_2en]^+$	neat PF_6^- salt	–	–	21584	21690	–	–	24800	–	[38, 62]
	neat $B\phi_4^-$ salt	–	–	21483	21725	–	–	25000	–	[38, 62]
$[Ir(ppy)_2bpy]^+$	$[Rh(ppy)_2bpy]PF_6$	–	–	21452	21465		–	23000[b]	21820	[45, 46]
$[Ir(ppy)_2en]^+$	$[Rh(ppy)_2en]PF_6$	–	–	20466[a]	20713[a]	–	–	22000[b]	–	[46]

[a] Maxima of the broad origin lines; [b] Estimated values; [c] In analogy to the related Ir^{3+} complexes the broad absorption features at the lowest energy were ascribed to the bpy.

Table 4. Calculated spin-orbit matrix elements (using Eq. (3)), calculated (using Eq. (2)) and observed Zero Field Splittings

Complex	$\langle ^1MLCT \vert H_{SO} \vert ^3LC \rangle$ [cm^{-1}]	ZFS_{calc} [cm^{-1}]	ZFS_{obs}: $\vert D\vert + \vert E\vert$ [cm^{-1}]	Method [Ref.]
$[Rh(thpy)_2bpy]^+$	41	0.04	0.144	ODMR [70]
$[Rh(ppy)_2bpy]^+$	34	0.04		
$[Ir(thpy)_2bpy]^+$	207	2.6	≈ 5	excit. [46]
$[Ir(thpy)_2en]^+$	310	13.6		
$[Ir(ppy)_2bpy]^+$	182	12.1	≈ 10	lumin./excit. [45]
$[Ir(ppy)_2en]^+$	201	17.5		

This would result in a dual emission from both thermally nonequilibrated excited states. The main characteristics of this luminescence would be non single exponential decay curves and luminescence bandshapes depending on the excitation wavelength. Among other complexes $[Ir(ppy)_2bpy]^+$ was reported to show such dual luminescence in a glassy matrix at 77 K [76, 77]. By wavelength selective excitation a broad and a structured band could be resolved, of which the one at higher energy was ascribed to a ^{3}MLCT (ppy^-) state and the broad band at lower energy to a ^{3}MLCT (bpy) state [77]. This explanation is not compatible with our spectroscopic results. In the previous chapters we showed that the ^{3}LC(ppy^-) and ^{3}MLCT (bpy) excited states of $[Ir(ppy)_2bpy]^+$ are close in energy and thermally equilibrated. We would ascribe the structured luminescence to a ^{3}LC(ppy^-) rather than to a ^{3}MLCT (ppy^-) luminescence. This is in fact supported by the transition dipole moments which were calculated from solvent induced band shifts [50]. The moments determined from absorption spectra were substantially smaller than the ones estimated from the luminescence spectra. There is no obvious reason why the ^{3}MLCT (ppy^-) transition should be weaker than the ^{3}MLCT (bpy) transition.

Let us try to give an alternative explanation for the phenomenon of the dual emission observed here. In the crystalline samples there is no dual emission. All the luminescence of $[Ir(ppy)_2bpy]^+$ originates from a ^{3}LC (ppy^-) lowest excited state indicating that there is an efficient relaxation of the higher lying ^{3}MLCT states (see Fig. 16). In solutions at room temperature there is no dual emission either, only luminescence from the ^{3}MLCT (bpy) excited state is observed. Its maximum is spectrally shifted below the ^{3}LC emission of the crystal. Also in this case the energy transfer between the excited states is evidently very fast. In a glassy matrix dual emission is observed, see Fig. 4. The solvatochromic and rigidochromic effects (see Sect. 2) indicate that, depending on the surrounding, the ^{3}MLCT state can be found at any energy within the shaded area in Fig. 16. The upper borderline of the shaded area corresponds to a crystalline environment, while the lower line represents a room temperature solution. The large distribution of sites in the glassy matrix covers a big part of this energy range, some of the sites providing a molecular surrounding with more crystalline character and others with more solution character. The energy of the ^{3}LC

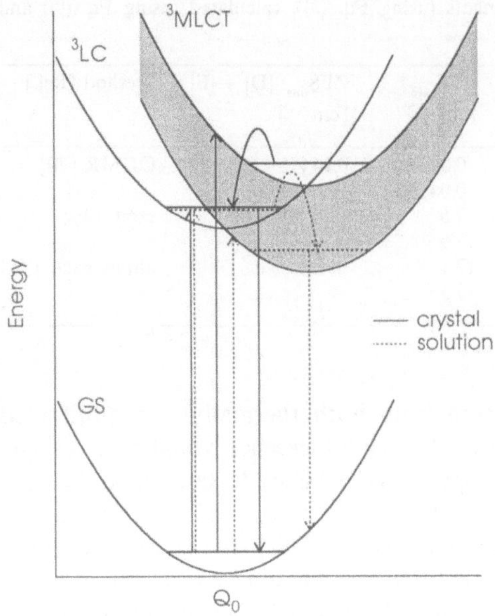

Fig. 16. Schematic configurational coordinate diagram of the ground and ^{3}LC and ^{3}MLCT excited states of the Ir^{3+} complexes. The *full* and *broken lines* refer to the state energies and relaxation pathways of the complex in a crystal or in solution, respectively. *Straight arrows* correspond to radiative and *curved arrows* to nonradiative relaxation processes. The *shaded area* indicates the range, in which the ^{3}MLCT state can be found, depending on the environment

excited state has a much smaller dependence on the environment. As a result, there are two subsets of molecules within the inhomogeneous distribution: one subset with a ^{3}LC and the other with a ^{3}MLCT first excited state. The ratio of the two subsets is temperature dependent, ^{3}MLCT dominating at high temperatures. The width of the distribution of sites does not have to span the full energy difference between the observed absorption maxima of the ^{3}LC and ^{3}MLCT transitions to produce the two subsets. The relevant energy difference is between the electronic origins of the two states which is much smaller due to the large Stokes shift for the MLCT state. It should be noted here that an analogous explanation for the dual emission from different *d–d* states of Cr^{3+} imbedded in inorganic glasses has been suggested [78].

Mixed ligand complexes offer a unique opportunity to elucidate the phenomenon of dual luminescence. Dual emission with different lifetimes from the two types of ligands has been postulated for the mixed-ligand complexes $[Rh(bpy)_x(phen)_{3-x}]^{3+}$, x = 1,2 [79, 80]. By using much more selective and highly resolved spectroscopic techniques on our mixed ligand title complexes, we find no evidence for this type of dual emission. The first two excited states, which are designated C and D in all our crystal spectra correspond to $^3\pi–\pi^*$ excitations localized on the two crystallographically inequivalent cyclometalating ligands. Nonradiative relaxation from D to C thus corresponds to a $^3\pi–\pi^*$ excitation energy transfer between these two ligands. Selective excitation of the higher energy line D exclusively leads to C luminescence in case of the $thpy^-$ complexes. In $[Ir(ppy)_2bpy]^+$ because of thermal population also luminescence from the close lying D line is observed. In $[Rh(ppy)_2bpy]PF_6$ no intrinsic luminescence from the origin line N of the bpy ligand is observed after excitation in the UV. All excitation is transformed to the ppy^- ligands and lower energetic traps resulting in pure C/D as well as trap luminescence. In $[Rh(thpy)_2bpy]^+$ a lower limit $k \geq 10^9\ s^{-1}$ for the transfer rate $D \rightarrow C$ was estimated [43] and a similar limit is assumed for the $N \rightarrow C/D$ transfer in $[Rh(ppy)_2bpy]^+$. In a recent study of the mixed ligand complexes $[Ir(bpy)_x(phen)_{3-x}]^{3+}$, $x = 2, 3$ Krausz et al. found no evidence for thermally non-equilibrated emission either [81].

In summary the most likely explanation for dual emission with different lifetimes in d^6 compounds is the presence of more than one luminescent species in the sample.

6 Conclusions

In our survey we have shown that with the use of different high-resolution spectroscopic methods a detailed view of the excited state properties of cyclometalated Rh^{3+} and Ir^{3+} complexes can be obtained. By following the excited state properties in a whole series of compounds it was found that the energetic order of the singlet excited states can be well explained by basic concepts such as a molecular orbital picture derived from σ-donor and π–acceptor strengths of the ligands, but for an explanation of the order of the lower lying triplet excited states singlet-triplet splittings and solvent effects have to be considered. In order to understand the fine details such as oscillator strengths, polarization properties, luminescence lifetimes and zero field splittings the mixing of singlet and triplet states due to spin-orbit coupling has to be taken into account. In the case of the Ir^{3+} complexes in particular the mixing of MLCT character into the ^3LC state has quite dramatic effects because of the large spin-orbit coupling constant of Ir^{3+}.

As borne out by the reference list a number of research groups have contributed to the understanding of the basic photophysical properties of cyclometalated Rh^{3+} and Ir^{3+} complexes over the years.

Acknowledgement. We thank Gabriela Frei for many helpful discussions. Financial support by the Swiss National Science Foundation is gratefully acknowledged.

7 References

1. Kalyanasundaram K (1982) Coord Chem Rev 46: 159
2. Krause RA (1987) Struct Bonding 67: 1
3. Juris A, Balzani V, Barigelleti F, Campagna S, Belser P, von Zelewsky A (1987) Coord Chem Rev 84: 85

4. Krausz E, Ferguson J (1989) Progr Inorg Chem 37: 293
5. Kirch M, Lehn JM, Sauvage JP (1979) Helv Chim Acta 62: 1345
6. Dehand J, Pfeffer M (1976) Coord Chem Rev 18: 323
7. Vancheesan S, Kuriacose JC (1983) J Sci Ind Res 42: 132
8. Constable EC (1984) Polyhedron 3: 1037
9. Newkome GR, Puckett WE, Gupta VG, Kiefer GE (1986) Chem Rev 86: 451
10. Omae I (1988) Coord Chem Rev 83: 137
11. Evans DW, Baker GR, Newkome GR (1989) Coord Chem Rev 93: 155
12. Ryabov AD (1990) Chem Rev 90: 403
13. Maestri M, Balzani V, Deuschel-Cornioley C, von Zelewsky A (1992) Adv Photochemistry 17: 1
14. Ryabov AD (1992) In: Williams AF, Floriani C, Merbach AE (eds) Pespectives in coordination chemistry. Helvetica Chimica Acta, Basel, p 271
15. Nonoyama M, Yamasaki K (1971) Inorg Nucl Chem Letters 7: 943
16. Nonoyama M (1974) Bull Chem Soc Jpn 47: 767
17. Nonoyama M (1979) Bull Chem Soc Jpn 52: 3749
18. Sprouse S, King KA, Spellane PJ, Watts RJ (1984) J Am Chem Soc 106: 6647
19. King KA, Finlayson MF, Spellane PJ, Watts RJ (1984) Sci Pap Inst Phys Chem Res 78: 97
20. Mäder U, Jenny T, von Zelewsky A (1986) Helv. Chim. Acta 69: 1085
21. Ohsawa Y, Sprouse S, King KA, DeArmond MK, Hanck KW, Watts RJ (1987) J Phys Chem 91: 1047
22. Garces FO, King KA, Watts RJ (1987) Inorg Chem 27: 3464
23. Garces FO, Watts RJ (1990) Inorg Chem 29: 583
24. van Diemen JH, Haasnoot JG, Hage R, Reedijk J, Vos JG, Wang R (1991) Inorg Chem 30: 4038
25. Maeder U, von Zelewsky A, Stoeckli-Evans H (1992) Helv Chim Acta 175: 1320
26. Colombo MG, Güdel HU (1993) Inorg Chem 32: 3081
27. Nonoyama M (1985) Polyhedron 4: 765
28. Constable EC, Leese TA, Tocher DA (1990) Polyhedron 9: 1613
29. Watts RJ, Harrington JS, Van Houten J (1977) J Am Chem Soc 99: 2179
30. Wickramasinghe WA, Bird PH, Serpone N (1981) J Chem Soc Chem Commun 1284
31. Spellane PJ, Watts RJ, Curtis C (1983) Inorg Chem 22: 4060
32. Nord G, Hazell AC, Hazell RG, Farver O (1983) Inorg Chem 22: 3429
33. King KA, Watts RJ (1987) Am Chem Soc 109: 1589
34. King KA, Spellane PJ, Watts RJ (1985) J Am Chem Soc 107: 1431
35. Dedeian K, Djurovich PI, Garces FO, Carlson G, Watts RJ (1991) Inorg Chem 30: 1685
36. Drews H, Müller J, Quao K (1992) In: Nitrogen ligands in organometallic chemistry and homogeneous catalysis. Euchem Conference, Alghero, Italy, May 10–15, 1992
37. Colombo MG, Brunold TC, Riedener T, Güdel HU, Förtsch M, Bürgi HB, Inorg Chem, in press.
38. Frei G, Zilian A, Raselli A, Güdel HU, Bürgi HB (1992) Inorg Chem 31: 4766
39. Maestri M, Sandrini D, Balzani V, Maeder U, von Zelewsky A (1987) Inorg Chem 26: 1323
40. Colombo MG, Zilian A, Güdel HU (1990) J Am Chem Soc 112: 4581
41. Colombo MG, Zilian A, Güdel HU (1991) J Lumin 48 & 49: 549
42. Zilian A, Güdel HU (1991) Coord Chem Rev 111: 33
43. Zilian A, Güdel HU (1992) Inorg Chem 31: 830
44. Ichimura K, Kobayashi T, King KA, Watts RJ (1987) J Phys Chem 91: 6104
45. Colombo MG, Hauser A, Güdel HU (1993) Inorg Chem 32: 3088
46. Colombo MG, unpublished results
47. McRae EG (1957) J Phys Chem 61: 562
48. Marcus RA (1963) J Chem Phys 39: 1734
49. Marcus RA (1965) J Chem Phys 43: 1261
50. Wilde AP, Watts RJ (1991) J Phys Chem 95: 622
51. Personov RI (1983) In: Agranovich VM, Hochstrasser RM (eds) Spectroscopy and excitation dynamics of condensed molecular systems. North Holland, Amsterdam, p 555
52. Verwey JWM, Imbusch GF, Blasse G (1989) J Phys Chem Solids 50: 813
53. Riesen H, Krausz ER (1993) Comments Inorg Chem, 14: 323
54. Riesen H, Krausz E, Puza M (1988) Chem Phys Lett 151: 65
55. Zilian A, Mäder U, von Zelewsky A, Güdel HU (1989) J Am Chem Soc 111: 3855
56. Zilian A, Colombo MG, Güdel HU (1990) J Lumin 45: 111
57. Zilian A, Frei G, Güdel HU, (1993) Chem Phys 173: 513
58. Gropper H, Dörr (1963) Ber Bun Ges 67: 46

59. DeArmond MK, Huang WL, Carlin CM (1979) Inorg Chem 18: 3388
60. Imbusch GF (1987) In: Lumb MD (ed) Luminescence spectroscopy. Academic, London, p 1
61. McGlynn SP, Azumi T, Kinoshita M (1969) In: Molecular spectroscopy of the triplet state. Prentice-Hall, Englewood Cliffs, p 17
62. Frei G (1990) Diploma thesis. University of Bern, Bern
63. Zilian A (unpublished results)
64. Komada Y, Yamauchi S, Hirota N (1986) J Phys Chem 90: 6425
65. Nakamoto K (1986) In: Infrared and Raman spectra of inorganic and coordination compounds, 4th edn. Wiley, New York, p 208
66. Strommen DP, Malick PK, Danzer GD, Lumpkin RS, Kincaid JR (1990) J Phys Chem 94: 1357
67. Small GJ (1983) In: Agranovich VM, Hochstrasser RM (eds) Spectroscopy and excitation dynamics of condensed molecular systems. North Holland, Amsterdam, p 515
68. Riesen H, Krausz E, Zilian A, Güdel HU (1991) Chem Phys Lett 182: 271
69. Zilian A, Güdel HU (1992) J Lumin 51: 237
70. Giesbergen CPM, Sitters R, Frei G, Zilian A, Güdel HU, Glasbeek M (1992) Chem Phys Lett 197: 451
71. Westra J, Glasbeek M (1991) Chem Phys Lett 180: 41
72. Riesen H, Krausz E (1992) J Lumin 53: 263
73. Demas JN, Crosby GA (1970) J Am Chem Soc 97: 7262
74. Demas JN, Crosby GA (1971) J Am Chem Soc 93: 2841
75. Watts RJ, White TP, Griffith BG (1975) J Am Chem Soc 97: 6914
76. King KA, Watts RJ (1986) J Am Chem Soc 109: 1589
77. Wilde AP, King KA, Watts RJ (1991) J Phys Chem 95: 629
78. Henderson B, Imbush GF (1989) In: Optical spectroscopy of inorganic solids, Oxford University Press, Oxford
79. Halper W, DeArmond MK (1972) J Lumin 5: 225
80. Crosby GA, Elfring WH Jr (1976) J Phys Chem 80: 2206
81. Krausz E, Higgins J, Riesen H, Inorg Chem (submitted)

Spectroscopic Manifestations of Potential Surface Coupling Along Normal Coordinates in Transition Metal Complexes

David Wexler[1], Jeffrey I. Zink[1] and Christian Reber[2]

[1] Department of Chemistry and Biochemistry, University of California, Los Angeles, California 90024, USA
[2] Department of Chemistry, University of Montreal, Montreal (Quebec), Canada H3C 3J7

Table of Contents

Topics in Current Chemistry, Vol. 171

The emission spectrum of hexafluoroacetylacetonatedimethylgold(III), Me_2Au(hfacac) at 10 K and the excitation spectrum of bishexafluoroacetylacetonatoplatinum(II), Pt(hfacac)$_2$ in a molecular beam contain vibronic structure with an intensity distribution that is indicative of coupling between normal coordinates. Good fits to the spectra are obtained when two totally symmetric coordinates are coupled. The effects of coordinate coupling on electronic spectra are calculated by using the split operator technique for numerical integration of the time-dependent Schrödinger equation and the time-dependent theory of electronic spectroscopy. Spectra resulting from coupled surfaces contain unusual intensity distributions in the vibronic structure. The spectra are calculated and the trends in the intensity distributions are analyzed in terms of the sign and the magnitude of the distortion and the type and the magnitude of the coupling.

1 Introduction

The intensity distribution of the vibronic structure in the electronic emission and absorption spectra of transition metal compounds can often be interpreted and calculated accurately by using harmonic uncoupled potential surfaces (i.e., independent vibrational coordinates and electronic states uncoupled from each other) [1, 2]. The spectra are frequently rich in vibronic structure because the ground and excited electronic state potential surfaces are displaced along many normal coordinates and because the displacements are large enough to give rise to long progressions [1]. A simple pattern of the progressions that can often occur is one in which a low frequency ($\sim 10^2$ cm^{-1}) metal-ligand mode and a higher frequency ($\sim 10^3$ cm^{-1}) ligand centered mode both have significant displacements. In this case the spectrum consists of a sequence of clusters of bands. Each cluster is a replica of the others and consists of a Poisson distribution of intensities in the low frequency mode. The clusters are separated by the wavenumber of the high frequency mode. The relative intensities of each of the clusters also consists of a Poisson distribution determined by the progression in the high frequency mode.

The emission spectrum of hexafluoroacetylacetonatodimethylgold(III), Me_2Au(hfacac) at 10 K appears at first glance to fit the above description [3]. There are clusters of bands whose individual peaks are separated by 262 cm^{-1}. These clusters are separated from each other by 1352 cm^{-1}. However, the clusters are not replicas of each other. Instead, the intensity distribution of peaks in each cluster changes; each cluster to lower energy of the previous one has an intensity distribution in which the lower energy peaks of the cluster gain intensity relative to the higher energy peaks. The "non-replica" pattern cannot be accounted for by harmonic potential surfaces and requires a surface where the two modes (both totally symmetric) are coupled.

The excitation spectrum of bis(hexafluoroacetylacetonato)platinum(II), $Pt(hfacac)_2$, in a supersonic molecular beam shows a similar absence of replicas [4]. There are clusters of bands whose individual peaks are separated by 65 cm^{-1}. These clusters are separated from each other by 1350 cm^{-1}. Again, the intensity distribution of peaks in each cluster changes; each cluster to higher energy of the previous one has an intensity distribution in which the lower energy peaks of the cluster gain intensity relative to the higher energy peaks. The "non-replica" pattern requires a surface where the two modes (both totally symmetric) are coupled. Spectroscopic methods for studying isolated molecules under collision free conditions in a supersonic molecular beam have recently been developed [5]. Under these conditions, excitation and dispersed luminescence spectra may contain highly resolved vibronic structure even when the molecules are large and the condensed media spectra are almost featureless [5, 6].

In this chapter the effects on emission and excitation spectra of coupling two normal modes in one electronic state are examined in detail. The dual emphases of this chapter are the physical meaning of the trends in the vibronic structure, and the fits of the experimental spectra. Both linear and quadratic coupling are discussed, but the major emphasis is on quadratic coupling because the electronic transition to this form of the surface gives the best fit to the experimental spectra. The spectra are calculated by using the time-dependent theory of electronic spectroscopy [7–9] and the split-operator method of Feit and Fleck [10–12] for numerically integrating the time-dependent Schrödinger equation. These methods provide physical insight into the relationships between the initial position of the wavepacket on the coupled surface, the shape of the surface, and the electronic spectrum. They also provide exact numerical solutions. The reasons for the non-replica form of the progressions, the trends in the relative intensities of the vibronic features, and the best fits to the experimental spectra are discussed.

2 Time-Dependent Theory

2.1 Harmonic Potentials

The theoretical background that will be needed to calculate the excited state distortions from electronic emission and absorption spectra is discussed in this section. We will use the time-dependent theory because it provides both a powerful quantitative calculational method and an intuitive physical picture [7–11]. In this section we will concentrate on the physical picture and on the ramifications of the theory.

A cross section along one normal mode of a multidimensional surface that will be used to discuss the theory for electronic emission spectroscopy is shown

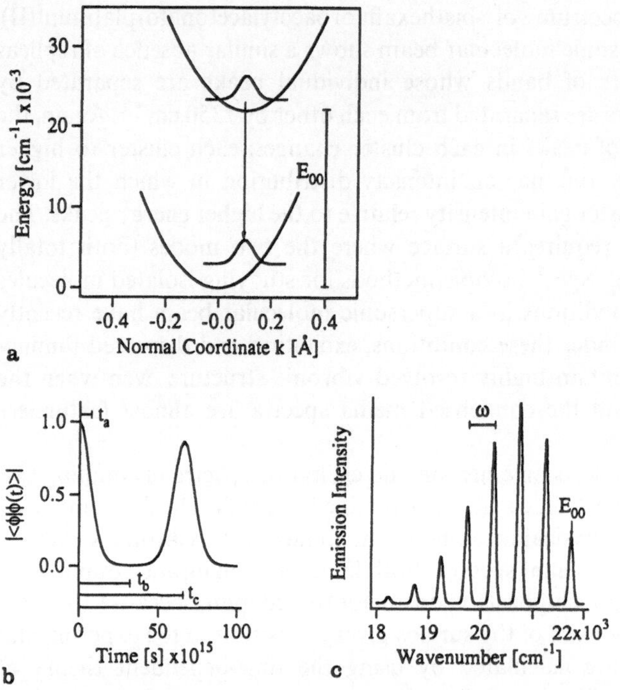

Fig. 1a–c. Illustration of the time dependent theory of emission spectroscopy for one-dimensional harmonic potential energy surfaces. **a** schematic view of the emission transition, **b** time dependence of the overlap $| < \phi | \phi(t) >$, **c** calculated emission spectrum

in Fig. 1a. The initial vibrational wavepacket, ϕ, makes a vertical transition and propagates on the ground state potential surface which, in general, is displaced relative to that of the excited state. The displaced wavepacket is not a stationary state and evolves according to the time-dependent Schrödinger equation. In electronic emission spectroscopy, the quantity of interest is the overlap of the initial wavepacket with the time-dependent wavepacket, $< \phi | \phi(t) >$.

The overlap for one specific normal mode (kth) has a simple form if it is assumed that (a) the force constants are the same in both ground and the excited states, (b) the potential surfaces are harmonic, (c) the transition dipole moment, μ, is constant and (d) the normal coordinates are not mixed in the excited state. With these assumptions, the overlaps, $< \phi | \phi(t) >$ in emission spectroscopy have the simple form

$$< \phi_k | \phi_k(t) > \; = \exp\left\{ -\frac{\Delta_k^2}{2}(1 - \exp(-i\omega_k t)) - \frac{i\omega_k t}{2} \right\} \tag{1}$$

where ω_k and Δ_k are respectively the vibrational frequency in cm^{-1} and the displacement of the kth normal mode.

The magnitude of the damped overlap, $| < \phi_k | \phi_k(t) > \exp(-\Gamma^2 t^2)|$ is plotted versus time in Fig. 1b. The overlap is a maximum at $t = 0$ and decreases as the wavepacket moves away from its initial position (t_a). It reaches a minimum when it is far away from the Franck-Condon region (t_b). At some time later t_c, the wavepacket may return to its initial position giving rise to a recurrence of the overlap. This time t_c corresponds to one vibrational period of the mode.

The electronic emission spectrum in the frequency domain is the Fourier transform of the overlap in the time domain. The emission spectrum is then given by

$$I(\omega) = C\omega^3 \int_{-\infty}^{+\infty} \exp(i\omega t)\left\{ < \phi|\phi(t) > \exp\left(- \Gamma^2 t^2 + \frac{iE_{00}}{\hbar} t\right)\right\} dt \quad (2)$$

where $I(\omega)$ is the intensity in photons per unit volume per unit time at frequency of emitted radiation, ω, C is a constant, and $< \phi|\phi(t) >$ is again the overlap of the initial wavepacket, $\phi = \phi(t = 0)$ with the time-dependent wavepacket $\phi(t)$. Γ is a phenomenological Gaussian damping factor and E_{00} is the energy of the origin of the electronic transition. Thus, the emission spectrum can be calculated when the frequencies ω_k and the displacements Δ_k of the normal modes are known. Figure 1c shows the Fourier transform of the overlap which is shown in Fig. 1b. In the frequency domain (emission spectrum), the vibronic spacing ω is equal to $2\pi/t_c$.

The physical picture of the dynamics of the wavepacket in the time domain described above provides insight into the absorption spectrum in the frequency domain. The width of the spectrum is determined by the initial decrease of the overlap which in turn is governed by the slope of steepest descent. The larger the distortion, the steeper the slope, the faster the movement of the wavepacket and the broader the spectrum. Any other features in the spectrum are governed by the overlap at longer times. For example, the vibronic spacing ω in the frequency domain is caused by the recurrence at time t_c corresponding to one vibrational period. The magnitude of the damping factor, to be discussed later, causes the magnitude of recurrences to decrease and gives the width of the individual vibronic bands.

In most transition metal and organometallic compounds, many modes are displaced. In the case of many displaced normal modes, the total overlap is

$$< \phi|\phi(t) > = \prod_k < \phi_k|\phi_k(t) > \quad (3)$$

Equations 1 and 3 are used to calculate the complete overlap using assumptions (a) though (d). It is given by

$$< \phi|\phi(t) > = \exp\sum_k\left[- \frac{\Delta_k^2}{2}(1 - \exp(- i\omega_k t)) - \frac{i\omega_k t}{2}\right] \quad (4)$$

The time-dependent theoretical treatment of the electronic absorption spectrum is very similar to that of the emission spectrum because the same two potential surfaces are involved. The principal difference is that the initial wavepacket starts on the ground state electronic surface and propagates on the excited electronic state surface. The overlap of the initial wavepacket with the time-dependent wavepacket is given by Eq (4) for the assumptions given at the beginning of this section. The absorption spectrum is given by [9]

$$I(\omega) = C\omega \int_{-\infty}^{+\infty} \exp(i\omega t)\left\{ < \phi|\phi(t) > \exp\left(-\Gamma^2 t^2 + \frac{iE_{00}}{\hbar} t \right)\right\} dt \qquad (5)$$

where all of the symbols are the same as those in Eq. (2). Note that for absorption, the intensity is proportional to the first power of the frequency times the Fourier transform of the time dependent overlap. Note also that the relationships between spectral features and the distortions are the same as those discussed for emission.

2.2 Coupled Normal Coordinates

When two or more normal coordinates are coupled, the wavepacket dynamics depends on all of the coupled coordinates simultaneously. Thus, the $\phi(t)$ cannot be written as the product of the $\phi(t)$'s for each coordinate computed individually and Eq. (3) cannot be used. Instead, the multidimensional wavepacket must be calculated by using the time dependent Schrödinger equation. In this section we show how to calculate the wavepacket for two coupled coordinates. The computational method discussed here removes all of the restrictive assumptions used in deriving Eq. (4). Any potential, including numerical potentials, can be used.

The time-dependence of the wavepacket evolving on any potential surface can be numerically determined by using the split operator technique of Feit and Fleck [10–15]. A good introductory overview of the method is given in Ref. [12]. We will discuss a potential in two coordinates because this example is relevant to the experimental spectra. The time-dependent Schrödinger equation in two coordinates Q_x and Q_y is

$$i\frac{\partial \phi}{\partial t} = -\frac{1}{2M}\nabla^2 \phi + V(Q_x,Q_y)\phi \qquad (6)$$

where $\nabla^2 = \partial^2/\partial x^2 + \partial^2/\partial y^2$ and $V(Q_x,Q_y)$ is the potential energy surface. The wavefunction at a time $t + \Delta t$ is [10]

$$\phi(t+\Delta t) = \exp\left(\left(\frac{i\Delta t}{4M}\right)\nabla^2\right)\exp(-i\Delta t V) \times \exp\left(\left(\frac{i\Delta t}{4M}\right)\nabla^2\right)\phi(Q_x,Q_y,t=0) + O[(\Delta t)^3]. \qquad (7)$$

The dominant error term is third order in Δt. The initial wavefunction $\phi(Q_x,Q_y,t)$ at $t = 0$ is normally the lowest energy eigenfunction of the initial state of the spectroscopic transition. The value of the wavefunction at incremental time intervals Δt is calculated by using Eq. (7) for each point on the (Q_x,Q_y) grid. The autocorrelation function is then calculated at each time interval and the resulting $\langle \phi | \phi(t) \rangle$ is Fourier transformed according to Eq. (2) to give the emission spectrum.

In the specific cases of the spectra of the metal compounds discussed in this chapter, a third mode is observed with no evidence of coupling to the Q_x and Q_y coordinates. In the framework of the time-dependent theory, the total autocorrelation function is the product of $\langle \phi | \phi(t) \rangle$ of the coupled coordinates discussed above and the $\langle \phi | \phi(t) \rangle$ from a separate calculation for the third mode.

The two most important choices that must be made in the numerical calculations are the size of the time steps and the size of the computational grid. The smaller the increment in the time steps and the smaller the spacing between the grid points the greater the accuracy in the calculation. General criteria for initial choices have been published [11, 12]. In the calculations reported here, the time increment is considered small enough and the total time large enough when the plot of the calculated spectrum does not distinguishably change when the increment is halved and the total time is doubled. A time increment $\Delta t = 0.60$ fs was suitable. The grid spacing is considered to be small enough when the plot of the calculated spectrum does not distinguishably change when the increment is halved. Grids of 128×128 (50 points/dimensionless unit along Q) defining the potential surfaces were generally suitable for the spectra calculated in this chapter. Typical calculations required about 150 seconds of CPU time on an IBM ES9000 computer.

3 Spectral Signatures of Mode Coupling

The form of the potential that will be used in this discussion is

$$V(Q_x,Q_y) = 0.5(k_x)Q_x^2 + 0.5(k_y)Q_y^2 + (k_{coupling})Q_x^u Q_y^v \tag{8}$$

The first two terms are harmonic potentials in coordinates Q_x and Q_y respectively. When $k_{coupling} = 0$, the spectra can be calculated by using Eq. (3). The simplest form of the coupling is linear, i.e. $u = v = 1$. This coupling is the well-known Duschinsky rotation [16–18]. The potential with $u = 1$ and $v = 2$ (or vice versa) corresponds to quadratic coupling. The experimental spectra are best described by this form of coupling. Contour plots of potentials corresponding to both types of coupling are shown in Fig. 2. We begin our discussion of the effects of coupling on electronic spectra with quadratic coupling.

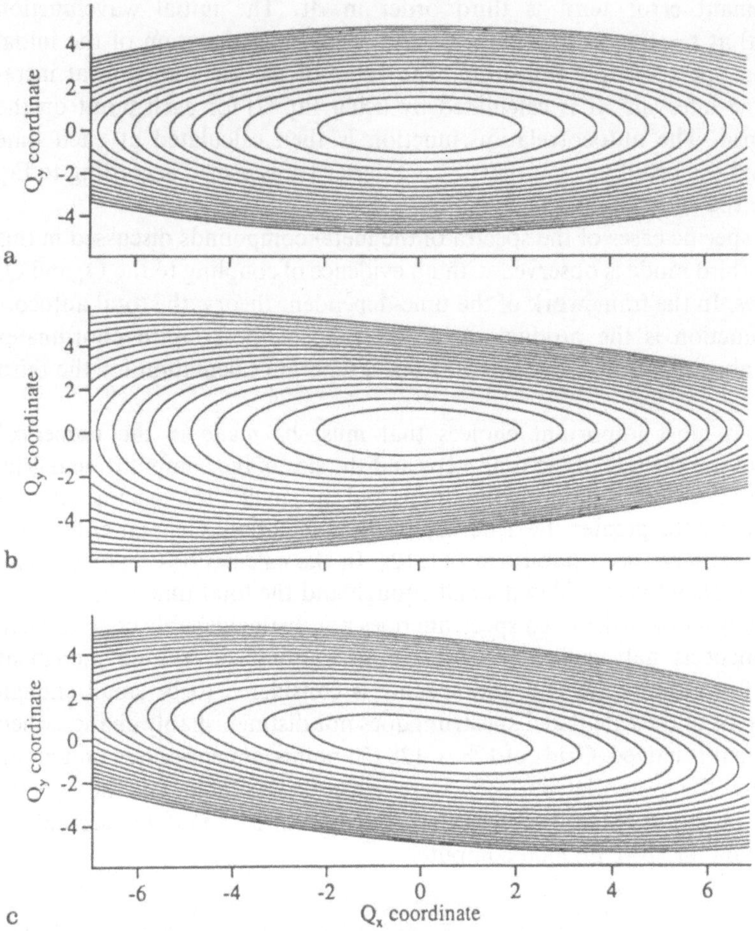

Fig. 2a–c. Two-dimensional potential energy surfaces. The Q_x coordinate corresponds to the low frequency mode, the Q_y coordinate corresponds to the high frequency mode. The forms of the coupling are: **a** $k_{xy} = 0$ (no coupling), **b** $k_{xy} = xy^2$ (quadratic coupling), and **c** $k_{xy} = xy$ (linear/Duschinsky coupling

3.1 Quadratic Coupling

The functional form of the potential is [19]

$$V(Q_x, Q_y) = 0.5(k_x)Q_x^2 + 0.5(k_y)Q_y^2 + (k_{xy})Q_xQ_y^2 \quad (9)$$

The coupling term that is used in this discussion is linear in the Q_x coordinate and quadratic in the Q_y, coordinate because this form best models the spectra presented in this chapter.

Contour plots of the uncoupled surface defined by $k_x = 262\ \mathrm{cm}^{-1}$, $k_y = 1352\ \mathrm{cm}^{-1}$ and $k_{xy} = 0$, and of the coupled surface having the same values of k_x and k_y but with $k_{xy} = 116.5\ cm^{-1}$ are shown in Figs. 2a and 2b respectively [20]. The coupled potential is symmetric about the Q_x axis. Therefore the sign of the displacement Δ of the initial wavepacket in the Q_y coordinate does not affect the calculated spectrum. However, the potential surface is asymmetric with respect to the Q_y coordinate axis and the calculated spectrum is dependent on the sign of the displacements in the Q_x coordinate. Another feature of the coupled surface that will play a major role in the following discussion is that the slope of the potential in the Q_y direction is dependent on the position along the Q_x coordinate. Starting at the origin and moving in the negative direction along the Q_x axis, the slope of the potential along the Q_y coordinate becomes smaller, i.e., the separation between the contour lines for a given ΔQ_y increases. Conversely, for increasing values of Q_x, the slope of the potential along the Q_y coordinate increases. These dependences of the slope along the Q_y coordinate on the value of Q_x have numerous effects on electronic spectra as will be seen in the following paragraphs.

3.1.1 Displacement in One Coordinate

The simplest starting point for examining the effects of potential surface coupling on the vibronic structure in electronic spectra involves displacement in only one coordinate. A displacement of the initial wave packet along the Q_x coordinate is chosen because it best illustrates the features necessary to fit the emission spectrum of Me_2Au(hfacac). The initial wavepacket is a harmonic oscillator function from a surface with $k_x = 262\ cm^{-1}$ and $k_y = 1352\ cm^{-1}$. The reference spectrum for the subsequent discussion is calculated by using a displacement of 1.44 (all displacements will be presented in dimensionless parameters) [20] along Q_x in the uncoupled potential surface (Fig. 3a). This magnitude of the distortion is chosen because it results in a vibronic progression similar to that observed in the spectrum of Me_2Au(hfacac). The spectra resulting from propagating the wavepacket on the coupled potential surface starting at $Q_x = +1.44$ and $Q_x = -1.44$ are shown in Figs. 3b and 3c.

The sign of the displacement along the Q_x coordinate in the coupled surface has a significant effect on the vibronic intensity distribution in the spectrum. With a positive displacement, the relative intensities of the vibronic bands with non-zero quantum numbers increase compared to the intensity of the origin band with quantum number zero. Thus the width of the progression is "broader" than that of the progression in the reference spectrum calculated by using the uncoupled surface and the same displacement (Figs. 3b and 3a respectively). In the broader spectrum, the vibronic intensity shifts to higher quantum number and the peaks to lower energy in the emission spectrum gain intensity. For example, in Fig. 3a, the peaks corresponding to $n = 0$ and $n = 1$ quanta in the low frequency mode and zero quanta in the high frequency mode are almost equal in height, but in Fig. 3b, the coupled spectrum, the $n = 1$ peak has gained intensity. In the remainder of this paper, the term "broader" will be used to refer to these relative changes in vibronic intensity caused by coupling

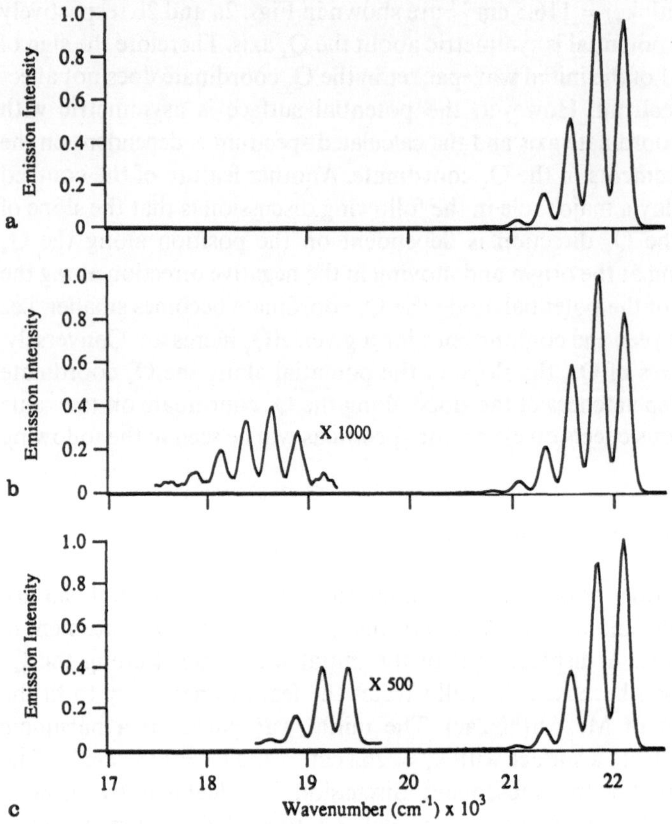

Fig. 3a–c. Calculated emission spectra with the initial wave packet displaced in the Q_x coordinate only. The spectra are calculated for: **a** the uncoupled potential surface (Fig. 2a) $k_{xy} = 0\ cm^{-1}$, $\Delta = 1.44$ (the reference spectrum); **b** the quadratically coupled potential surface (Fig. 2b) $k_{xy} = 116.5\ cm^{-1}$ and $\Delta = 1.44$; and **c** the quadratically coupled potential surface (Fig. 2b) $k_{xy} = 116.5\ cm^{-1}$ and $\Delta = -1.44$

and the progressions with n quanta in the Q_x dimension and m quanta in the Q_y dimension will be written as (n,m).

When the wavepacket is displaced in the negative direction, the opposite effect occurs. The width of the progression in the new spectrum (Fig. 3c) is "narrower" than that in the reference spectrum, i.e., vibronic peaks with higher quantum numbers are less intense. For example, in Fig. 3a, the highest peak in the (n, 0) progression is the one with n = 1, but in the coupled spectrum (Fig. 3c), the n = 1 peak is less intense and the highest peak is the one with n = 0. In the remainder of this paper, the term "narrower" will be used to refer to these relative changes in vibronic intensity caused by coupling.

A second feature of note is the weak spectroscopic feature consisting of a vibronic progression in the low frequency mode built on one quantum of the high frequency mode (Figs. 3b and 3c). Vibronic bands from the high frequency mode (corresponding to the Q_y coordinate) appear in the spectrum even though the wavepacket is not displaced along that coordinate. This intensity is a direct result of the coupling and is absent in the spectrum from the uncoupled surface.

The features discussed above can be readily interpreted in terms of the time-dependence of the autocorrelation function $<\phi|\phi(t)>$. The most important factor governing the changes in the widths of the spectra for the positive versus negative displacement is the initial decrease in $<\phi|\phi(t)>$. When transforming from the time domain to the frequency domain, a slow decrease in $<\phi|\phi(t)>$ corresponds to a narrow progression and a fast decrease in $<\phi|\phi(t)>$ corresponds to a broader progression.

The decrease in $<\phi|\phi(t)>$ depends on the slope of the potential surface at the point at which the wavepacket is initially placed. The slope of the potential in the Q_y direction is steeper on the positive Q_x side of the surface than on the negative Q_x side. When the slope in the Q_y dimension is large, the Q_y part of the two-dimensional wavepacket will rapidly change its shape and $<\phi|\phi(t)>$ will decrease rapidly. Therefore, for a positive Q_x displacement in the coupled potential, the autocorrelation function will decrease more rapidly than it would for a negative displacement.

The magnitudes of $<\phi|\phi(t)>$ versus time are shown in Fig. 4. The autocorrelation function for the positive displacement along Q_x in the coupled potential (lowest curve in Fig. 4), starts out at 1 and drops to 0 over a shorter period of time than in the uncoupled potential (middle line). The Fourier transforms of these $<\phi|\phi(t)>$ give the spectra shown in Fig. 3. This reasoning explains why a positive displacement results in a broader progression and a negative displacement results in a narrower progression.

The physical insight obtained from the time-domain point of view allows simple qualitative predictions about the widths of the progressions to be made. Two potential surfaces (such as the coupled and uncoupled surfaces in Fig. 2) can be compared in terms of the slope of steepest descent at the position of the initial wavepacket at $t = 0$. The steeper the slope, the faster the initial decrease of $<\phi|\phi(t)>$ in the time domain and the broader the spectrum in the frequency domain. This insight is important in developing a strategy to fit experimental spectra. In the example above, a vibronic progression in an experimental spectrum that is broader than that expected from a harmonic potential surface (Fig. 3a) requires that the wavepacket be displaced in the positive Q_x direction in the coupled potential surface.

The appearance of a vibronic band corresponding to the high frequency mode (even though the wavepacket is not displaced along the Q_y coordinate) can also be qualitatively understood in terms of wavepacket dynamics on the coupled potential surface. The explanation is closely related that used to explain a progression in an undisplaced coordinate when there is a change in the force constant between the initial and final surfaces: the initial wavepacket is not an eigenfunction of the surface and evolves with time. Thus, a wavepacket that is an eigenfunction of the cross section of the surface at $Q_x = 0$ is not an eigenfunction of a cross section at $Q_x \neq 0$ and will develop in time. These oscillations in the

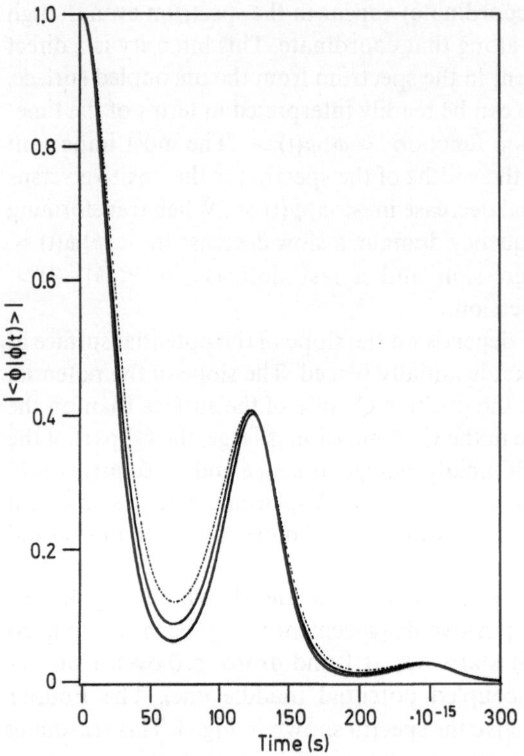

Fig. 4. Autocorrelation functions plotted versus time for the spectra described in Fig. 3. The autocorrelation function shown by the *middle curve* corresponds to the reference spectrum (Fig. 3a), $k_{xy} = 0\ cm^{-1}$, the autocorrelation function shown by the *lowest curve* corresponds to a positive displacement in the quadratically coupled potential surface, and the autocorrelation function shown by the *top curve* corresponds to a negative displacement in the quadratically coupled potential surface

time domain correspond to vibronic bands in the frequency domain with energy spacings corresponding to the Q_y (high frequency) coordinate. These oscillations in $< \phi|\phi(t) >$ are too small to be seen in the plots shown in Fig. 4.

On the basis of the above discussion, it is obvious that increasing the magnitude of the distortion along the Q_x coordinate broadens the progression independent of the sign of the distortion. The larger the distortion, the steeper the slope, the faster the decrease in $< \phi|\phi(t) >$ and the broader the progression.

3.1.2 Displacements in Two Coordinates

A more general picture of the effects of coupling of two coordinates on the electronic emission spectrum is obtained when the initial wavepacket is displaced along both of the coupled modes. This situation is required to interpret the experimental spectrum of Me_2Au(hfacac). In order to illustrate the trends, the same potential surfaces that were discussed above are used again, but the initial wavepacket is displaced by 1.44 along the Q_x coordinate and by 1.52 along the Q_y coordinate. (These values give the best fit to the experimental spectrum, vide infra.) The spectrum resulting from the uncoupled potential surface, shown in Fig. 5a, is a useful reference in order to understand the changes brought about by the coupling. It consists of a series of progressions, each identical to that in Fig. 3a, separated by 1352 cm^{-1}, the wavenumber of the high frequency mode. The progressions in the low frequency mode, (n,0), (n,1), (n,2), etc. are all replicas of one another because the wavepacket experiences a slope in the Q_y direction that is independent of the Q_x coordinate and a slope in the Q_x direction that is independent of the Q_y coordinate. The relative intensities of these replicas are determined by the intensities (Poisson distribution) of the progression in the high frequency mode [e.g., (0,m)]. The identical spectrum can

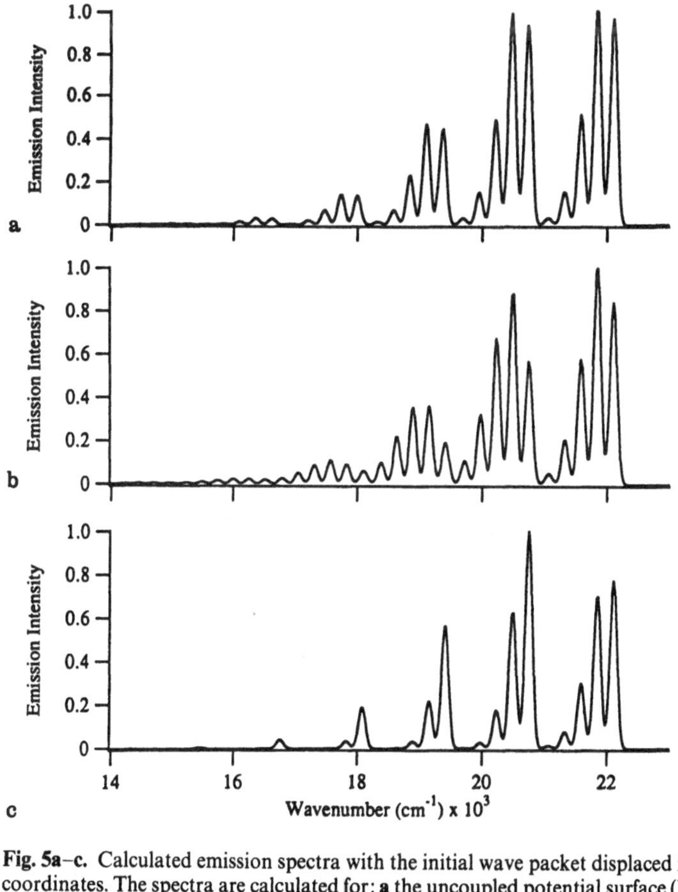

Fig. 5a–c. Calculated emission spectra with the initial wave packet displaced in both the Q_x and Q_y coordinates. The spectra are calculated for: **a** the uncoupled potential surface (Fig. 2a) $k_{xy} = 0\ cm^{-1}$, $\Delta_x = 1.44$, and $\Delta_y = 1.53$ (the reference spectrum); **b** the quadratically coupled potential surface (Fig. 2b) $k_{xy} = 116.5\ cm^{-1}$, $\Delta_x = 1.44$, and $\Delta_y = 1.53$; and **c** the quadratically coupled potential surface (Fig. 2b) $k_{xy} = 116.5\ cm^{-1}$, $\Delta_x = -1.44$, and $\Delta_y = 1.53$

be calculated by carrying out two separate one-dimensional calculations, multiplying the $<\phi|\phi(t)>$, and Fourier transforming the result.

The spectra resulting from displacing the initial wavepacket along the positive Q_y coordinate on the *coupled* potential surface and positively and negatively along the Q_x coordinate are shown in Figs. 5b and 5c, respectively (The sign of the displacement along the Q_y coordinate is not critical because the potential is symmetric about the Q_x axis). In the case of a positive displacement of the wavepacket along the Q_x coordinate (Fig. 5b), the low frequency progressions become broader with increasing quantum number of the high frequency mode. The progressions are not replicas of each other. For example, in Fig. 5b, the broadening of the low frequency progression can be observed by the increase in the intensity of the peaks corresponding to higher quantum number within each successive progression. In the (n,0) progression the most intense peak is the n = 1 peak, but in the (n,3) progression the n = 2 peak is the most intense.

For the negatively displaced case (Fig. 5c), the low frequency progressions are again not replicas of each other, but become narrower. The narrowing of the progressions can be observed in the decrease in intensity of peaks corresponding to larger quantum numbers when going from the (n,0) progression to the (n,1) and (n,2) and (n,3) etc. progressions. In the (n,0) progression, the n = 3 peak has significant intensity, but by the (n,2) progression the n = 3 peak has negligible intensity.

The fact that the sequences of progressions in the low frequency mode are not replicas of each other (i.e., the broadening or narrowing of the progressions as a function of the quantum number of the high frequency mode) is explained by the slope of the potential surface in a manner similar to that used in the one coordinate displacement case. The most important additional factor is that the displacement in the Q_y dimension places the initial wavepacket on a part of the surface with an even steeper slope in the Q_y dimension and thus causes a longer progression in the high frequency mode. The total width of the spectrum, dominated by the progression in the high frequency mode, is broader. For example, Figs. 3c and 5c show that the spectral change resulting from a change in the displacement from 0.0 to 1.53 along the high frequency Q_y coordinate is broadening of the spectrum. Figure 3c shows one large progression (n,0) and one small progression (n,1) whereas Fig. 5c shows the appearance of many quanta of the high frequency mode (n,2), (n,3), etc. In general, the larger the displacement in any coordinate, positive or negative, the broader the width of the progression arising from wavepacket dynamics in that coordinate.

3.1.3 Change in Force Constant

Another factor that can affect the width of the spectrum is the difference between the vibrational frequencies in the excited and ground electronic states. The trends in the spectral changes can be readily explained from the time-dependent

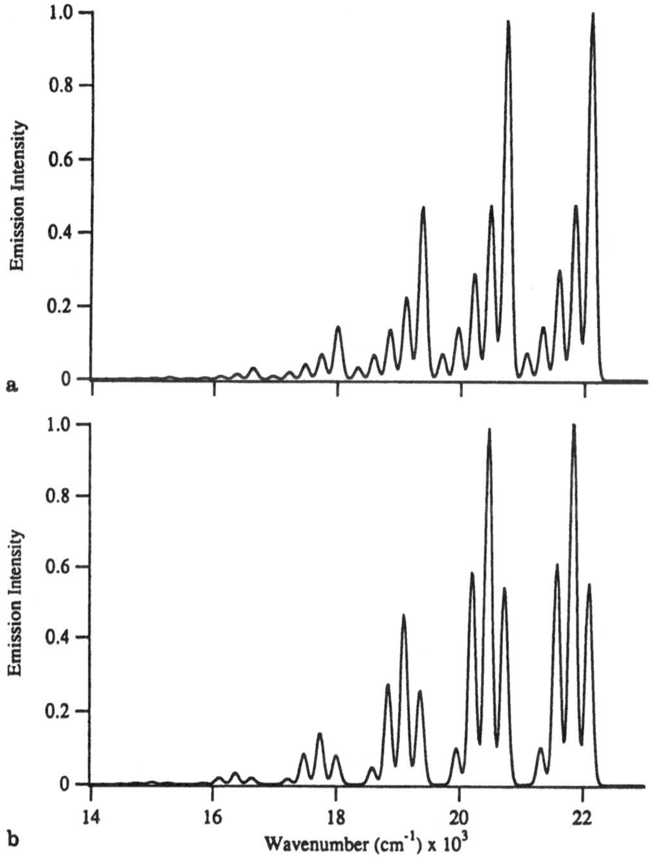

Fig. 6a, b. Calculated emission spectra showing the effects of changing the force constant of the initial wave packet placed on the quadratically coupled surface with $\Delta_x = 1.44$, and $\Delta_y = 1.53$. The excited state force constants are: **a** 540 cm^{-1} and **b** 135 cm^{-1}

point of view. Figures 6a and 6b show the effects of an increase and a decrease in the low frequency force constant on an emission spectrum. In Fig. 6a, the low frequency force constant of the initial wavepacket is increased from 262 cm^{-1} to 540 cm^{-1}. The spatial extent of the wavepacket in the Q_x coordinate becomes narrower. This narrower wavepacket yields a spectrum where the low frequency progression is narrower compared to the progression resulting from the propagation of the 262 cm^{-1} wavepacket. The opposite effect is observed, Fig. 6b, when the force constant is decreased from 262 cm^{-1} to 135 cm^{-1}.

The above effect is explained by the spatial extent of the initial wavepacket on the potential surface. A larger wavepacket (smaller force constant) covers more of the potential, experiences a larger slope in the Q_x dimension, and therefore increases the width of the low frequency progressions because the initial decrease in the overlap is fast. Conversely, a smaller wavepacket (larger force constant) covers less of the potential, experiences a smaller slope in the Q_x

dimension and therefore results in slower decrease of the overlap and narrower low frequency progressions.

3.2 Linear (Duschinsky) Coupling

The simplest form of coupling between two normal coordinates is linear, i.e. $(k_{xy})Q_xQ_y$. This form of the coupling is not able to account for the quantitative vibronic intensities in the emission spectrum of Me_2Au(hfacac). However, spectra resulting from linearly-coupled potential surfaces also show series of vibronic progressions that are not replicas of each other. A brief examination of the effects of linear coupling are included here in order to illustrate what spectral features can arise.

The two-dimensional potential including linear coupling is given by

$$V(Q_x,Q_y) = 0.5(k_x)Q_x^2 + 0.5(k_y)Q_y^2 + (k_{xy})Q_xQ_y \tag{10}$$

The force constants k_x and k_y that are used in this example are the same as those in the example of quadratic coupling discussed above, but k_{xy} is 262 cm^{-1} and has a significantly different meaning. This functional form of coupling is equivalent to a Duschinsky rotation [16–18] with a rotation angle of 12.85°. A contour plot of the rotated potential surface is shown in Fig. 2c. In order to illustrate the effects of displacing the initial wavepacket in both dimensions, displacements in the Q_x (low frequency) coordinate of 1.57 and in the Q_y (high frequency) coordinate of 1.5 are used. (These displacements are equal to those for the harmonic potential model best fit to the experimental emission spectrum.) The spectra resulting from displacements in two coordinates in both the uncoupled and the coupled potential surfaces are shown in Fig. 7. Figures 7b and 7c are spectra resulting from positive and negative displacements in the Q_x coordinate respectively.

For a positive Q_x distortion, the intensity distributions of the progressions in the low-frequency mode are not replicas of one another, but become broader as the quantum number of the high frequency mode increases. The increases in breadth in the low frequency progressions can be seen be comparing the (n,0) progression to the (n,1) progression in Fig. 7a to the (0,n) progression in Fig. 7b. Two additional effects are also observed when the coupled spectrum is compared to the uncoupled potential spectrum. First, the high frequency progression is broader in the coupled surface spectrum than in the reference spectrum, i.e., the total spectrum is broader. This increase in breadth is seen by comparing the relative intensities of the (n,0) peaks in Fig. 7a to those in Fig. 7b. Second, a given low frequency progression is narrower in the coupled spectrum than that in the reference spectrum as can be seen by comparing the (0,n) progression in Fig. 7a to the (0,n) progression in Fig. 7b. This result means that a given progression in the coupled surface spectrum is narrower than the corresponding progression in the harmonic model reference spectrum, despite the fact that the ensuing progressions become broader.

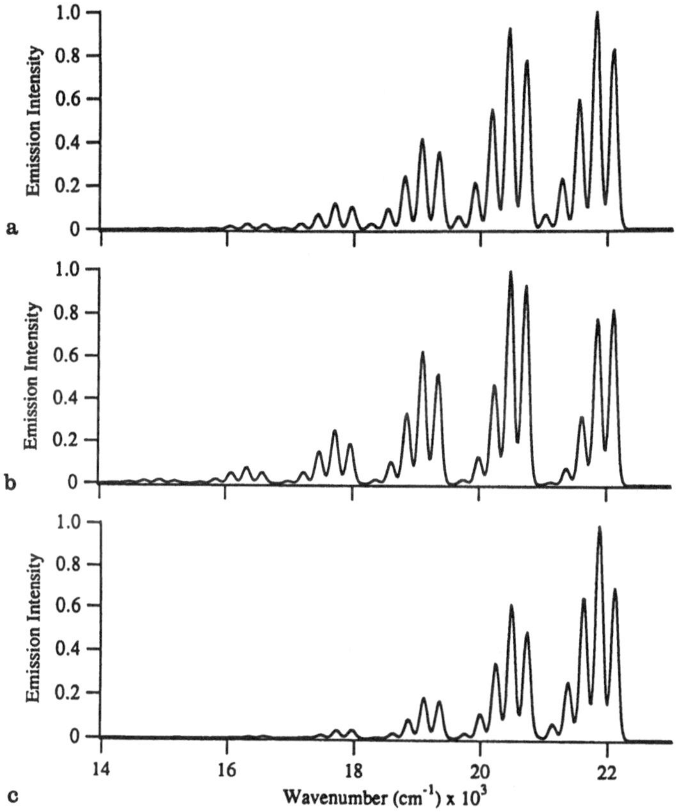

Fig. 7a–c. Calculated emission spectra illustrating the effects of linear (Duschinsky) coupling and displacing the initial wave packet in both the Q_x and Q_y coordinates. The spectra are calculated with **a** $k_{xy} = 0\ \mathrm{cm}^{-1}$, $\Delta_x = 1.57$, and $\Delta_y = 1.5$ (the reference spectrum); **b** $k_{xy} = 262\ \mathrm{cm}^{-1}$, $\Delta_x = 1.57$, and $\Delta_y = 1.5$; and **c** $k_{xy} = 262\ \mathrm{cm}^{-1}$, $\Delta_x = -1.57$, and $\Delta_y = 1.5$

For a negative distortion in the Q_x coordinate, the opposite effects occur. The low frequency progressions became narrower within the coupled surface spectrum. The high frequency progression is also narrower in the coupled surface spectrum compared to the reference spectrum. For example, in Fig. 7a (the spectrum from the uncoupled potential), there is significant intensity in 5 quanta whereas in Fig. 7c (the spectrum form the coupled potential), there is significant intensity in only 4 quanta.

4 Interpretation of the Emission Spectrum of Me_2Au(hfacac)[3]

The emission spectrum of Me_2Au(hfacac) shows that the potential surface is coupled along two normal coordinates. The clusters of bands separated by

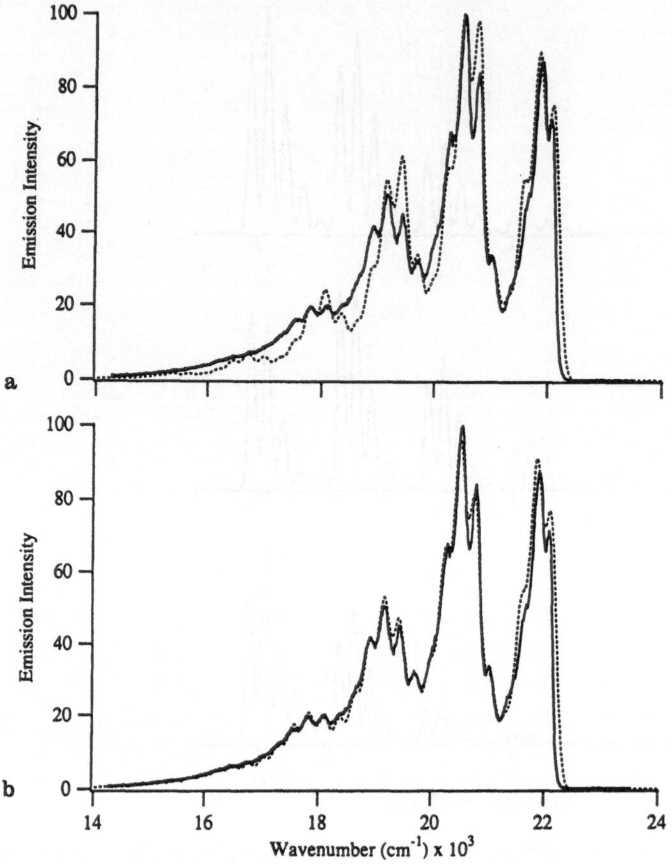

Fig. 8a, b. Comparisons of the calculated (*broken lines*) and experimental (*solid lines*) emission spectra of frozen Me_2Au(hfacac) at 15 K. **a** The "best fit" using the uncoupled harmonic potential surface. **b** The best fit using the quadratically coupled potential surface

262 cm^{-1} are not replicas of each other. The spectrum is shown in Fig. 8 (solid line).

The emission is assigned as a Π^* to Π ligand centered transition, consistent with the assignments for previously reported metal-hexafluoroacetylacetonate complexes [21–23]. Three ring modes are active in the spectrum. The two main progressions in the spectrum are created by a low energy mode of 262 cm^{-1} and a high energy mode of 1352 cm^{-1}. A third mode, having an energy spacing of 1050 cm^{-1}, appears as a single peak on the high energy side of each quanta of the 1352 cm^{-1} mode. The 262 cm^{-1} mode is assigned as ring deformation, the 1352 cm^{-1} mode as primarily a C–C ring stretch coupled to a C–CF_3 stretch, and the 1050 cm^{-1} mode as a CF_3 rock [24–26] All of the active modes involve motions of the ligand atoms, consistent with the ligand centered Π^* to Π assignment for the electronic transition.

A simple starting point for fitting an experimental spectrum is the use of uncoupled harmonic potential surfaces. The theoretical spectra are calculated by using Eqs. (2, 6, 7 and 9). The adjustable parameters in the fitting procedure are the displacements Δ along the normal coordinates and the force constants k_x, k_y, and k_{xy}. The k's defining the potential surface are obtained from the emission spectrum and/or from vibrational spectra. The values of E_{00} and Γ that appear in Eq. (2) are obtained from the experimental spectrum.

The intensities of the vibronic bands observed in the experimental spectrum cannot be fit by using an uncoupled harmonic potential surface. The harmonic potential "best fit" to the (n,0) progression between 21 000 and 23 000 cm^{-1} is obtained by using $\Delta_x = 1.57$, $\Delta_y = 1.5$, $k_x = 1352\,cm^{-1}$, $k_y = 262\,cm^{-1}$. The "best fit" in Fig. 8a includes the 1050 cm^{-1} mode with $\Delta = 0.87$. Although a good fit to the (n,0) progression is obtained, the fits of the (n,1), (n,2) etc. progressions are poor. The reason for the poor match is the increase of the width of the low frequency progression with increasing quantum number of the high frequency mode, (i.e., the progressions are not replicas of each other).

When fitting a spectrum that shows obvious effects of coupling, the simplest form of the coupling that can be used is linear (Duschinsky rotation). Based on the above discussion, in order to fit the experimental spectrum in Fig. 1, the wavepacket would have to be placed in a steep region of the potential. This requirement is met if the wavepacket is displaced in both the positive Q_x and positive Q_y direction or in both the negative Q_x and negative Q_y direction. Attempts to fit the experimental spectrum by using linear coupling and the above displacements were not as successful as those with quadratic coupling because the vibronic intensities and the spacings between the peaks could not be matched simultaneously.

The fact that the progressions become broader requires that the initial wavepacket be displaced to a part of the quadratic coupled potential surface where the slope is steeper than that of the harmonic surface. The required change is similar to that illustrated in Fig. 5b. Thus, to obtain a good fit the wavepacket must be displaced in a positive direction along the Q_x coordinate on the coupled surface. The coupled surface requires the use of the coupling parameter k_{xy}. It is important to note that k_x and k_y alone do not define the vibrational frequencies of the fundamentals in the coupled surface. The vibrational frequencies and the spacings between the vibronic peaks are determined by the values of k_x, k_y and k_{xy}. Thus, when k_{xy} is varied in the fitting procedure, the values of k_x and k_y must also be changed such that the eigenvalues of the coupled surface and hence the spacings of the vibronic peaks match those determined experimentally from either the electronic spectrum or from a vibrational spectrum. Increasing the value of k_{xy} makes the positive Q_x part of the potential surface steeper in Q_y and the negative Q_x part flatter. Thus for a given Δ_x and Δ_y, increasing k_{xy} magnifies the effects of broadening or narrowing discussed above. The Δ's and the k's are sequentially changed in directions leading to increasingly between matches between the calculated and experimental spectra.

The best fit between the experimental and calculated spectra is shown in Fig. 8b. The values of the parameters for the transition to ground state potential are: $E_{00} = 22910\ cm^{-1}$, $\Gamma = 74\ cm^{-1}$, $\Delta_x = 1.44$, $\Delta_y = 1.52$, $k_x = 258\ cm^{-1}$, $k_y = 1354\ cm^{-1}$, $k_{xy} = 116.5\ cm^{-1}$. The initial wavepacket parameters for the fit are as follows: $k_{x\,ini.wvpkt} = 270\ cm^{-1}$ and $k_{y\,ini.wvpkt} = 1352\ cm^{-1}$. The uncoupled mode ($1050\ cm^{-1}$) parameters are: $k_{ini.wvpkt} = 1100\ cm^{-1}$, $k_{potential} = 1050\ cm^{-1}$, and $\Delta = 0.87$. The value of E_{00} was obtained from the energy of the highest energy (0,0) peak in the experimental emission spectrum and the phenomenological damping factor (Γ) was determined by the line widths in the experimental emission spectrum. The other parameters were adjusted, as described above, so that the vibronic intensities and peak placements matched the Me_2Au(hfacac) emission spectrum.

The calculated vibronic intensities are extremely sensitive to small changes in Δ. Changes in Δ on the order of a few percent have large effects on the spectrum. The best fit uses a displacement $\Delta = 1.44$ in the low frequency mode and a displacement $\Delta = 1.52$ in the high frequency mode. If the parameters are changed to $\Delta_x = 1.42$ and $\Delta_y = 1.48$, for example, a significantly different spectrum is obtained. The spectrum resulting from these small changes in Δ (1.6% in Δ_x and 3% in Δ_y) has intensity changes in the vibronic structure ranging from 6–10%. The relative heights of the high frequency progression show the largest changes in peak intensity.

The changes in the shape of the molecule based on the parameters used in the best fit are shown schematically in Fig. 9. In the one-electron picture, the transition resulting in the emission removes an electron from Π-antibonding orbital on the ligand ring and places it in a Π-bonding orbital. This transition results in an increase in the Π-bonding. Based on this reasoning, the high frequency progression is interpreted to result from a decrease in the carbon-carbon bond length. (The sign of the change cannot be obtained from the fit

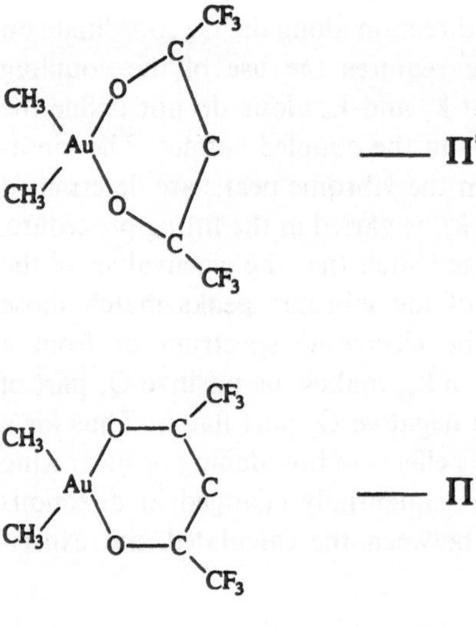

Fig. 9. Schematic diagram of the molecular structures of Me_2Au(hfacac) illustrating the change between the excited electronic state structure (*top*) and the ground state structure (*bottom*)

because this coordinate appears as Q_y^2 in the potential.) As shown in Fig. 9, the shortening of the C–C bonds will result in an increase in the O–C–C bond angles and a decrease in the Au–O–C bond angles. The non-zero coupling means that the force constant of the high frequency mode depends on the ring bond angles. As the O–C–C bond angles increase and the Au–O–C bond angles decrease, the C–C force constant decreases.

5 Interpretation of the Excitation Spectra of $Pt(hfacac)_2$ in a Molecular Beam[4]

The spectrum of $Pt(hfacac)_2$ also shows evidence of mode coupling. In the spectrum there are clusters of peaks separated by 1350 cm^{-1} that are not replicas of each other. The spectrum to be interpreted is the excitation spectrum. In this case, an excitation spectrum is the same as an absorption spectrum. Therefore we will use Eq. (5) in conjunction with Eqs. (6 and 7) to calculate the spectra.

The excitation spectrum of $Pt(hfacac)_2$ in the supersonic molecular beam is shown in Fig. 10. Strong signals were recorded between 22000 cm^{-1} and 26000 cm^{-1}. The dyes which were used to span this relatively large range and

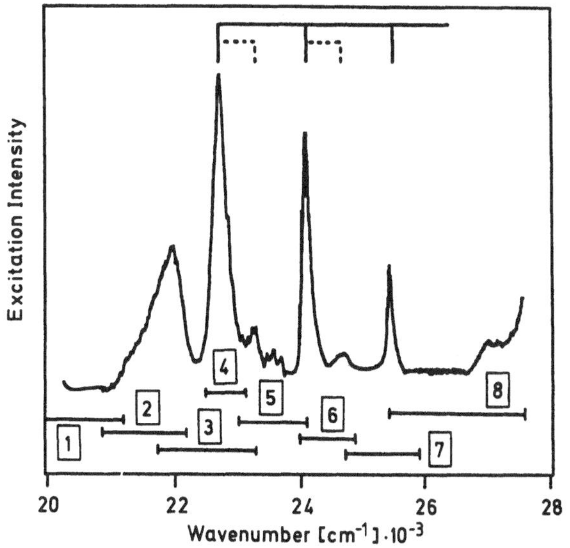

Fig. 10. Excitation spectrum of $Pt(hfacac)_2$ in a molecular beam. The *solid* and *dotted lines* on top of the spectrum indicate the progressions in the 1350 cm^{-1} and 597 cm^{-1} modes, respectively. The wavenumber regions of the dyes used to obtain the spectrum are shown below the spectrum. The dyes are: (*1*) Coumarin 480, (*2*) Coumarin 460, (*3*) Coumarin 440, (*4*) POPOP, (*5*) Bis-MSB, (*6*) DPS, (*7*) alpha-NPO, (*8*) BBQ.

the specific regions accessible with the individual dyes are also shown. The lowest energy feature is a broad band with very little resolved structure. Its onset occurs at about 21 000 cm^{-1} and its maximum is at 29 000 cm^{-1}. The higher energy region of the spectrum is dominated by a set of three sharper bands with maxima at 22 700 cm^{-1}, 23 985 cm^{-1}, and 25 400 cm^{-1}. These bands, with an average separation of 1350 cm^{-1}, contain resolvable fine structure. Finally, smaller features containing fine structure are observed between the three sharp, intense bands.

The magnified spectrum in the region between 22 500 cm^{-1} and 23 500 cm^{-1} containing the first of the intense, sharp bands as well as some of the smaller features is shown in Fig. 11. The vibronic fine structure on the most intense band is evident in the expanded spectrum. These vibronic bands are not well resolved, but a regular spacing between the shoulders of 65 cm^{-1} is found. The vibronic structure in the least intense band in Fig. 11 is even less well resolved, but the same energy spacing is evident.

A partial list of the band energies, organized in a manner appropriate for the following discussion, is given in Table 1. The three most important energy spacings are the 1350 cm^{-1} separation between the sharp, intense features, the 597 cm^{-1} separation between the sharp features and the first peak to the higher energy side of the sharp features, and the 65 cm^{-1} average separation between the shoulders on these peaks.

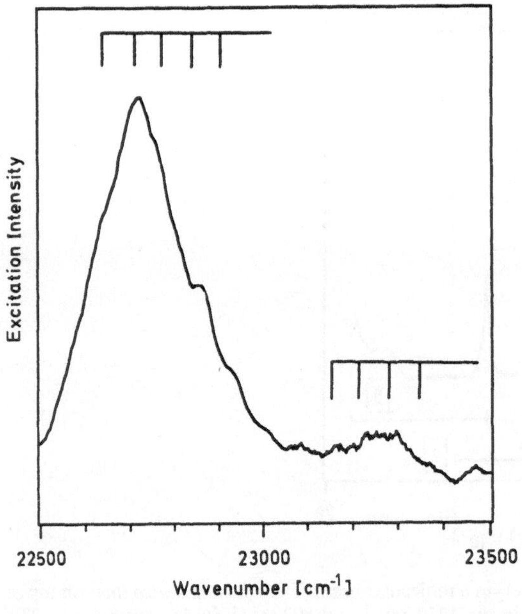

Fig. 11. Expanded view of the excitation spectrum of $Pt(hfacac)_2$ in a molecular beam in the region between 22 500 cm^{-1} and 23 500 cm^{-1}. The progression in the 65 cm^{-1} mode is indicated on the top

Table 1. Selected transition energies and assignments for the excitation spectra of $Pt(hfacac)_2$ in a molecular beam. Assignments refer to the numbers in the lefthand column

No.	Transition energy [cm^{-1}]	Assignment
1	22613	Electronic origin
2	22700	(1) + $\hbar\omega_{lig}$
3	22771	(1) + $2\hbar\omega_{lig}$
4	22843	(1) + $3\hbar\omega_{lig}$
5	22921	(1) + $4\hbar\omega_{lig}$
6	23220	(1) + $\hbar\omega_{PtO}$
7	23985	(1) + $\hbar\omega_{CO}$
8	24055	(7) + $\hbar\omega_{lig}$
9	24120	(7) + $2\hbar\omega_{lig}$
10	24645	(7) + $\hbar\omega_{PtO}$
11	25400	(1) + $2\hbar\omega_{CO}$
12	25455	(11) + $\hbar\omega_{lig}$

5.1 Assignments of the Excitation Bands

The bands observed in the excitation spectrum of $Pt(hfacac)_2$ in the molecular beam are assigned to transitions to two electronic states; a low lying ligand field excited state and a slightly higher energy ligand to metal charge transfer (LMCT) excited state. The band with its peak maximum at 21900 cm^{-1} and weak features to higher energy are assigned to the ligand field excited state. The weak features arise from vibronic structure. The prominent series of sharp intense bands at 22700, 23985 and 25400 cm^{-1} are assigned to vibronic components of the LMCT state. The weak bands at 597 cm^{-1} to higher energy from each of these bands are assigned to additional vibronic components of the LMCT state. The basis of these assignments are the molar absorptivities [27–30], and previously reported photoelectron spectroscopic results and molecular orbital calculations [31, 32].

The major experimental basis for assigning the bands to ligand field and charge transfer transitions is the molar absorptivity measured in the solution spectrum. The lowest energy band observed in the solution spectrum has a molar absorptivity on the order of 10^2 M^{-1} cm^{-1}. This value is typical for spin-allowed ligand field transitions in square planar platinum(II) complexes [27–30]. The next higher energy band in the solution spectrum at 23000 cm^{-1}, has a molar absorptivity of $> 10^3$ which is typical for a charge transfer transition. Similar bands, a weak feature at about 20000 cm^{-1} and a stronger feature at 25000 cm^{-1}, have been reported in the absorption spectrum obtained from a hot gas [33].

The assignment of the intense band in the beam to a LMCT transfer is supported by molecular orbital calculations [31]. Pseudo-potential ab initio calculations indicated that the highest occupied molecular orbital in $Pt(acac)_2$ is mainly ligand π antibonding in character. The results of the calculation compared well with photoelectron spectroscopic data [32]. Because the absorption spectrum of the acac complex is very similar to that of the hfacac complex investigated here, it is reasonable to assume that the assignment is the same in both complexes. Thus the intense band arises from a transition from the π ligand orbital to either an orbital primarily metal d in character (LMCT) or an orbital mainly ligand π antibonding in character (ligand π-π^*). Previous studies have shown that the ligand centered π-π^* transition in acac complexes occurs at higher energies in the region of 35000 cm^{-1} [34]. We thus assign the state at about 24000 cm^{-1} to LMCT.

5.2 Assignment of the Vibronic Structure

The vibronic peaks in the molecular beam excitation spectrum are assigned by comparison with the electronic ground state vibrations of the molecules and of analagous bis(acetylacetonato) complexes. The three prominent peaks separated by 1350 cm^{-1} are assigned to a progression in the totally symmetric CO stretching mode. The 1603 cm^{-1} mode in the Raman spectrum of $Pt(hfacac)_2$ and the 1565 cm^{-1} mode of $Pt(acac)_2$ arise from the totally symmetric CO stretch [35]. This mode is observed between 1520 cm^{-1} and 1600 cm^{-1} in a variety of divalent transition metal acetylacetonato complexes [36]. The decreased frequency in the excited electronic state relative to that in the ground state is expected due to the reduced CO bond order in the emitting state.

The small peaks that are found 597 cm^{-1} to higher energy from the sharp peaks in the molecular beam spectrum are assigned to a quantum of the PtO totally symmetric stretch. In the Raman spectrum of $Pt(hfacac)_2$ and $Pt(acac)_2$ the PtO stretch occurs at 635 cm^{-1} and 485 cm^{-1}, respectively [35]. It occurs between 420 cm^{-1} and 500 cm^{-1} in a series of metal acetylacetonates complexes [36]. The band in the 23000 cm^{-1} region of the molecular beam spectrum is overlapped by components of the ligand field band.

The structure on the above peaks with a separation of 65 cm^{-1} arises from a progression in a totally symmetric ring deformation mode. Low frequency modes in the 100–200 cm^{-1} range of the IR spectrum of $Pt(hfacac)_2$ and $Pt(acac)_2$ have been observed and attributed to ring deformation modes but not completely assigned [36, 37].

The spacings corresponding to the above assignments are shown in Figs. 10 and 11. The LMCT spectrum is dominated by progressions in these three modes. Distortions in these modes are expected in a transition which involves the π system on the ligand and a metal *d* orbital involved in Pt–O bonding. The magnitudes of the distortion and the intensity of specific features of the vibronic structure are analyzed in the following sections.

5.3 Normal-Mode Coupling and Calculation of the Excitation Spectrum

Starting with $\phi(t = 0)$, the evolution of ϕ with time can be calculated to high accuracy. The most important components that determine the excited state potential are the excited state distortions, i.e. the position of the potential minimum along the configurational coordinates of the vibrational modes involved. Only modes with a displacement in the excited state contribute to vibronic structure in an allowed transition and therefore only these modes need to be considered in our calculation.

The modes involved in the vibronic structure of the LMCT transition of $Pt(hfacac)_2$ are the C–O stretching, the Pt–O stretching and a ligand ring bending vibration. The simplest trial potential V is harmonic along the configuration coordinates Q of all three modes:

$$V = V_{harm} = \frac{1}{2}\left\{k_{CO}(Q_{CO} + \Delta_{CO})^2 + k_{PtO}(Q_{PtO} + \Delta_{PtO})^2 + k_{lig}(Q_{lig} + \Delta_{lig})^2\right\} \tag{11}$$

Both Q_i and Δ_i are in dimensionless units and therefore the force constants k_i are equal to the excited state vibrational energies in wavenumbers. The latter are obtained from the spectrum (Fig. 10 and Table 1). The Δ_i denote the excited state distortions and are the parameters to be determined. The values of both E_{00} (22 613 cm^{-1}) and Γ(19 cm^{-1}) are obtained from the spectrum in Fig. 10. The initial wavepacket ϕ is calculated by using the literature values for the ground state vibrational energies along the 3 modes.

The best fit spectrum calculated by using the harmonic, uncoupled potential with $\Delta_{CO} = 1.18$, $\Delta_{PtO} = 0.48$ and $\Delta_{lig} = 2.18$ gives good overall agreement in view of the simplicity of the model. The main progression in the 1350 cm^{-1} mode with its low energy fine structure is well reproduced. The major discrepancy between the experimental and calculated LMCT spectra involves the widths of the transitions forming the 1350 cm^{-1} progression. The experimental widths are 326 cm^{-1}, 152 cm^{-1} and 102 cm^{-1} for the bands in order of increasing energy. These widths are governed by the progression in the 65 cm^{-1} ligand ring bending mode which is superimposed on every member of the 1350 cm^{-1} progression in the C–O stretching vibration. The uncoupled harmonic calculation gives a *constant* width of 202 cm^{-1} for all the members of the main progression. The uncoupled potential surface model is therefore not able to account for the observed trend in these widths. The changes in the widths are evidence for coupling between the normal coordinates in the excited state.

The form of the coupling that best fits the experimental features is quadratic, Eq. (9), where Q_x is Q_{lig} and Q_y is Q_{CO}. This coupled potential surface includes the coupling constant as an adjustable parameter. It is shown for $k_{xy} = 0.25$ in Fig. 12. The time evolution of the wavepacket ϕ on this surface is no longer independent along Q_{CO} and Q_{lig} and therefore a calculation on the full two

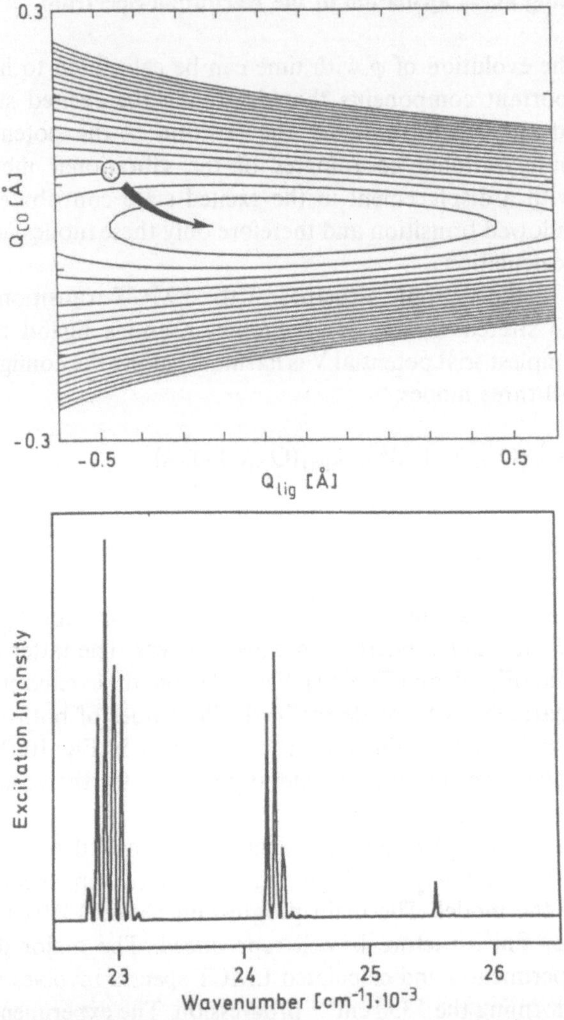

a

b

Fig. 12a. Excited state potential surface illustrating coupling between the 1350 cm^{-1} and 100 cm^{-1} normal modes. The potential surface was calculated by using Eq. (5). The starting position of the wavepacket is shown by the *dot*. **b** The excitation spectrum calculated by using this surface. The different band widths are shown

dimensional surface is needed for the two coupled modes. Equation (3) can no longer be applied to this situation. Because no experimental evidence of coupling is observed for the Pt–O stretching mode, its autocorrelation is calculated independently and multiplied to the autocorrelation for the two coupled modes to obtain the full spectrum.

The excitation spectrum calculated with the two dimensional potential surface in Fig. 12a is shown in Fig. 12b. The starting position of the moving wavepacket is shown as a black dot in Fig. 12a. The calculated spectrum in Fig. 12b shows the pronounced effects of the coupling: the widths of the bands narrow with increasing transition energy and show the same trend as that observed in the experiment. In this pedagogical calculation a value of 5 cm^{-1} was chosen for Γ in order to clearly illustrate the changes in the low-energy progression on the three peaks separated by 1350 cm^{-1}. For the situation in Fig. 12b and in the spectrum in Fig. 10, where three members of the high energy progression are observable, the widths decrease if the wavepacket starts on the shallow side of the coupled potential and increase if the wavepacket starts on the steep side of the coupled potential.

The calculated spectrum with the coupled harmonic excited state potential, Eq. (9) is compared to the experimental spectrum in Fig. 13. The agreement between the calculation and the experiment is significantly better than with the uncoupled harmonic potential, Eq. (11). The observed trend of narrowing of the members of the main progression is reproduced by starting the wavepacket on the shallow side of the potential as illustrated in Fig. 12a. The remaining small discrepancies are most likely due to the simple functional form of the coupling term and the limitation of the coupling to only two of the three vibrational modes. The values of the parameters for the coupled harmonic potential, Eq. (9) are $\Delta_{CO} = 1.04$ (− 11% change from the uncoupled model), $\Delta_{lig} = 2.16$ (− 1% change), $\Delta_{PtO} = 0.48$ (unchanged) and $k_{xy} = 0.11$. The best fit values of Δ obtained with the coupled and uncoupled potential surfaces are therefore very similar, although the effect of coupling on the spectrum is pronounced.

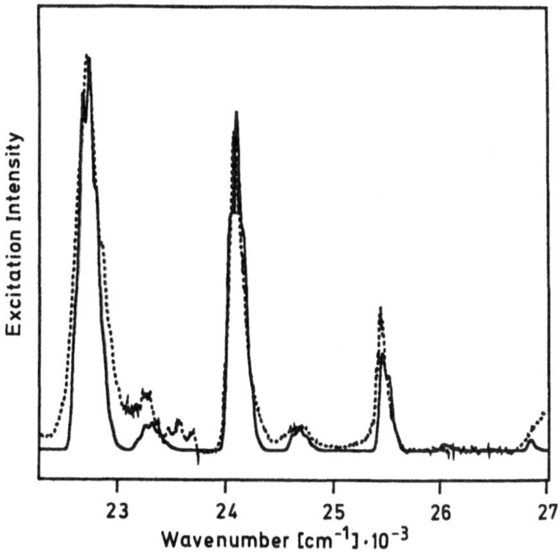

Fig. 13. Comparison of the experimental spectrum (*dotted line*) with that calculated by using the excited state potential surface defined in Eq. (5) with coupling between the two modes (*solid line*), as described in Sect. 5.3

5.4 Excited State Geometry

The values of Δ obtained by fitting the calculated spectrum to the experimental spectrum determine the minimum of the potential surface in configuration coordinate space and hence the bond length and bond angle changes which occur when the molecule is excited. In order to convert the dimensionless distortions into bond length and bond angle changes, a full normal coordinate analysis is required. Because such an analysis has not been completed, a few simple assumptions will be used to give an estimate of the distortions. For both stretching vibrations, the mass was assumed to be 16 g/mol, and for the ligand ring bending mode it was assumed to be 28 g/mol (one CO fragment). The relationships between the bond length changes δ_i, the bond angle changes $\delta\alpha_i$, and the normal coordinates ΔQ_i are approximated to be

$$\Delta Q_{CO} = \frac{1}{2}(\delta_{CO1} + \delta_{CO2} + \delta_{CO3} + \delta_{CO4}) \tag{12}$$

$$\Delta Q_{PtO} = \frac{1}{2}(\delta_{PtO1} + \delta_{PtO2} + \delta_{PtO3} + \delta_{PtO4}) \tag{13}$$

$$\Delta Q_{lig} = \frac{1}{2}(\delta\alpha_{CP-Pt-CO1} + \delta\alpha_{CO-Pt-CO2} + \delta\alpha_{CO-C-CO1} + \delta\alpha_{CO-C-CO2}) \tag{14}$$

The numerical values obtained from Eq. (12–14) and the Δ's obtained from the fit using the coupled potential are $\delta_{CO} = 0.023$ Å, $\delta_{PtO} = 0.011$ Å and $\delta\alpha = 2.0°$. The values obtained by using the uncoupled potentials are $\delta_{CO} = 0.021$ Å, $\delta_{PtO} = 0.011$ Å and $\delta\alpha = 2.0°$. The differences between the distortions calculated with the coupled and uncoupled models are very small, probably smaller than the uncertainties introduced by using the approximations in Eqs. (12–14) for the normal coordinates. The excited state structure of $Pt(hfacac)_2$ determined from the excitation spectrum in the molecular beam is shown schematically in Fig. 14.

Two aspects of the distortions deserve further comment. First, it is surprising that the distortions calculated by using the coupled potential surface differ only slightly from those calculated by using the simple uncoupled surface model. The coupling which is involved is small, but it causes obvious changes in the spectrum. The changing widths in the jet spectrum clearly show that coupling occurs. However, even the simple model provides a very good estimate of the excited state geometry. Second, the distortions that occur are those which are expected in a LMCT transition. The major changes are expected to be in the metal-ligand Pt–O bond and in the C–O and C–C bonds in the ring, as are observed. The largest change occurs in the C–O bond on the ligand.

For this molecule the almost featureless crystal spectra do not allow such a detailed analysis of the excited state geometry and potential surface to be made.

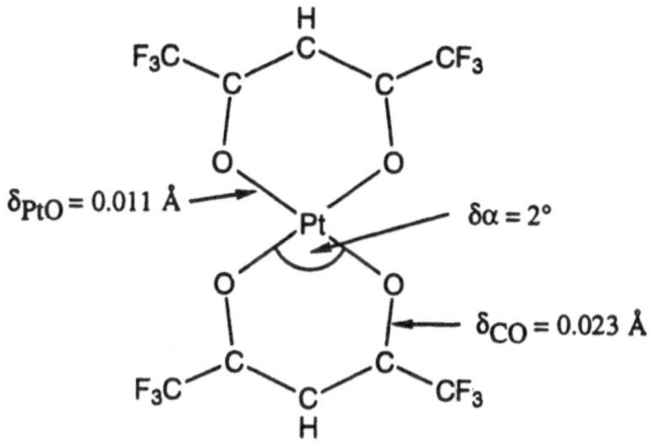

Fig. 14. Excited state distortions of $Pt(hfacac)_2$ in the molecular beam

Excitation spectroscopy in a molecular beam is a powerful technique leading to more insight into the excited state properties of transition metal compounds.

6 Summary

The vibronic structure in the low temperature emission spectrum of $Me_2Au(hfacac)$ and in the excitation spectrum of $Pt(hfacac)_2$ was quantitatively calculated and interpreted by using a potential surface with coupling between two totally symmetric normal coordinates and the time-dependent theory of electronic spectroscopy. Insight into the trends in the intensities in vibronic structure caused by changing the magnitudes and signs of the displacement on coupled surfaces was presented and strategies for fitting experimental spectra were discussed.

When vibronic structure is resolved in electronic spectra, it can often be interpreted by using the conventional Franck–Condon analysis for uncoupled potential surfaces. However, in transition metal complexes where potential surfaces are frequently displaced along many modes involve nuclear motions with one or more atoms in common, normal coordinates of the same symmetry may be coupled. We describe several of the spectral signatures of such coupling. When coupling exists, the time-dependent theory provides a powerful method of calculating and interpreting the spectra. The physical picture based on wave-packet motion on potential surfaces provides insight into the intensity distribution.

Acknowledgements. This chapter is dedicated to the memory of Prof. Dr. Günter Gliemann. This work was made possibly by a grant from the National Science

Foundation (NSF CHE 91-06471) and a grant for research collaboration by the Ministère de l'enseignement supérieur du Québec.

References

1. Zink JI, Kim Shin K-S (1991) Molecular distortions in excited electronic states determined from electronic and resonance Raman spectroscopy, In: Volman DH, Hammond GS, Neckers DC (eds) Advances in Photochemistry, vol. 16. John Wiley & Sons, New York and reference therein
2. Yersin H, Otto H, Zink JI, Gliemann G (1980) J Am Chem Soc 102: 951
3. Wexler D, Zink JI (1993) J Phys Chem 97.
4. Reber C, Zink JI (1992) Inorg Chem 31: 2786
5. Smalley RE, Wharton L and Levy DH (1977) Accounts of Chem Res 10: 139
6. Subbi J (1984) Chem Phys Letters 1: 109
7. Heller EJ (1981) Acc Chem Res 14: 368
8. Heller EJ (1978) J Chem Phys 68: 3891
9. Heller EJ (1978) J Chem Phys 68: 2066
10. Feit MD, Fleit JA, Steiger A (1982) J Comp Phys 47: 412
11. Kosloff D, Kosloff R (1983) J Comp Phys 52: 35
12. For an introductory overview see Tanner JJ (1990) J Chem Ed 67: 917
13. Reber C, Zink JI (1992) J Chem Phys 96: 2681
14. Reber C, Zink JI (1992) Comments Inorg Chem 13: 177
15. Rever C, Zink JI (1991) Coord Chem Rev 111: 1
16. Duschinsky F (1937) Acta Physicochim URSS 7: 551
17. Small GJ (1971) J Chem Phys 54: 3300
18. Subbi J (1988) J Chem Phys 122: 157
19. The force constant k_x is given by $k_x = 4\pi^2 M_x(\hbar\omega_x)^2$ when M is the mass and $\hbar\omega_x$ is the wavenumber of the fundamental. For convenience, the k's are expressed in units of cm^{-1}. The coupling constant force constant k_{xy} is given by

$$k_{xy} = \frac{\sqrt{c^5}}{\hbar^3}\sqrt{m_x(\hbar\omega_x)}\,[m_y(\hbar\omega_y)](\hbar\omega_{xy}).$$

20. The minimum of the excited state potential surface is displaced from that of the ground state potential surface by $\Delta = Q(\text{excited state}) - Q(\text{ground state})$. The formula to convert the dimensionless displacement Δ into Å is

$$\delta = \sqrt{\frac{6.023 \times 10^{23}}{M} \times \frac{\hbar}{2\pi c\omega}} \times 10^8 \times \Delta$$

where M is the mass involved in the vibration in the units of gram atomic weight (e.g. C = 12 g), ω is the wavenumber of the vibrational mode in cm^{-1}, $\hbar = h/2\pi$ where h is Planck's constant in $g\,cm^2\,s^{-1}$, c is the speed of light in $cm\,s^{-1}$, δ is the displacement in Å, and Δ is the dimensionless displacement. When values of M are chosen, the k's and k_{xy} must be scaled according to the expressions in footnote 19.
21. Armendarez PX, Forster LS (1964) J Chem Phys 40: 273
22. Halverson F, Brinen JS, Leto JR (1964) J Chem Phys 40: 2790
23. Gafney HD, Lintuedt RL, Jaworiwsky IS (1970) Inorg Chem 9: 1738
24. Behnke GT, Nakamoto K (1967) Inorg Chem 6: 433
25. Miles MG, Glass GE, Tobias RS (1966) J Am Chem Soc 88: 5738
26. Nakamoto K (1986) Infrared and Raman spectra of inorganic and coordination compounds; 4th edn. Wiley, New York, p 260
27. Chang TH, Zink JI (1985) Inorg Chem 24: 4499
28. Chang TH, Zink JI (1986) Inorg Chem 25: 2736
29. Fanwick PE, Martin DS (1973) Inorg Chem 12: 24

30. Jorgensen CK (1962) Absorption spectra and chemical bonding in complexes. Pergamon Press, Oxford, chap 6
31. Di Bella D, Fragala I, Granozzi G (1986) Inorg Chem 25: 3997
32. Fragala I, Costanzo LL, Ciliberto E, Condorelli G, D'Arrigo C (1980) Inorg Chim Acta 40: 15
33. Mingxin Q, Nonot R, van den Bergh H (1984) Scientia Sinica 27: 531
34. Lintvedt RL, Kernitsky LK (1970) Inorg Chem 9: 491
35. Kradenov KV, Kolesov BA, Zharkova GI, Isakova VG (1988) Zhurnal Struk Khimii 31: 56
36. Nakamoto K, McCarthy PJ, Martell A (1961) J Am Chem Soc 83: 1272
37. Oglezneva IM, Igumenov IK (1984) Koord Khim 10: 313.

Author Index Volumes 151-171

Author Index Vols. 26-50 see Vol. 50
Author Index Vols. 51-100 see Vol. 100
Author Index Vols. 101-150 see Vol. 150

The volume numbers are printed in italics